Geothermal Energy Development

PROBLEMS AND PROSPECTS IN THE IMPERIAL VALLEY OF CALIFORNIA

Edgar W. Butler and James B. Pick

University of California
Riverside, California

Plenum Press • New York and London

Library of Congress Cataloging in Publication Data

Butler, Edgar W.
  Geothermal energy development.

  Bibliography: p.
  Includes index.
  1. Geothermal engineering—California—Imperial County. I. Pick, James B. II. Title.
  TJ280.7.B87   333.3'815'0979499   81-23402
  ISBN 0-306-40772-8   AACR2

© 1982 Plenum Press, New York
A Division of Plenum Publishing Corporation
233 Spring Street, New York, N.Y. 10013

All rights reserved

No part of this book may be reproduced, stored in a retrieval system, or transmitted,
in any form or by any means, electronic, mechanical, photocopying, microfilming,
recording, or otherwise, without written permission from the Publisher

Printed in the United States of America

To my parents
Helen Block Pick and Grant Julius Pick
J.P.

To Pattie, Brian, and Tracey Butler
E.B.

Foreword

What are the effects on an isolated region when an entirely new and major energy resource is developed to commercial proportions? What happens to the population, the economy, the environment, the community, and societal relations? How does the government framework respond, the family structure adapt, the economy expand, and life styles change under the impact of new forces which hold a promise of much benefit and a risk of adverse consequences?

Imperial County, California, has a population of less than 90,000 people. This population has been exceptionally stable for years, centered as it is in an agricultural and recreational framework. The county is somewhat cut off from other areas by geographic barriers of mountains and desert, by state and natural boundaries, and is the most remote of all 58 counties of California from the state capitol, Sacramento.

In the decade of the 1950s, geographical explorations for oil revealed some anomalous structures underlying the desert and agricultural areas in Imperial County. These, when drilled, seemed to be oil-less and hot, and so lacked attractiveness to petroleum wildcatters.

In the decade of the 1960s, Dr. Robert Rex from the University of California, Riverside, and other scientists developed more completely the geological features of the region, including explorations of high-flow zones associated with the multiple faulting of the area, and for circulation of underground water which might be the conduit for major heat flows from deep magmatic intrusions to relatively reachable depths. The discovery of wells yielding very hot and very concentrated brines near Niland baffled commercial development for either energy or for minerals, but fired the imagination of pioneer geothermal developers.

Meanwhile, geothermal heat as a source of electric power and also by-product heat gained increasing attention world-wide as a petroleum shortage loomed. Pioneer developments in Italy and New Zealand were followed by development of the world's largest geothermal-electric generating complex in the pure steam area of The Geysers, California. Then, Mexico pioneered a mixed phase flashed steam power plant at Cerro Prieto, just a few miles south of Imperial County.

In the 1970s came a series of geothermal discoveries in Imperial County: Heber, North Brawley, and East Mesa, along with the earlier Niland discovery—all of which have attracted commercial developers seeking to generate electric power for export to metropolitan San Diego and Los Angeles. Many apparently successful wells have been drilled and tested, and the first power-generating facilities are now under construction.

The people of Imperial County have regarded these developments with optimistic interest and with equanimity. They appear to look hopefully for new economic development which may be overlaid on the agricultural base to provide jobs for young people who would otherwise migrate out. They do not want the attractive agricultural lifestyle to be destroyed, yet they would encourage more widely available power sources for air conditioning and heating in the severe desert climate, and for intensive agricultural and industrial applications. They are open to change and to economic development provided it does not impair the quality of air, water, and land; strict environmentalists and "no-growth" advocates are not widely found among the population.

Under these circumstances, local government has moved rationally and with dispatch to make possible the development of the geothermal resource. The county has handled zoning and permitting procedures in a manner often held up as a fine example of protection of public interest with minimal red tape.

So, Imperial Valley now faces a construction phase in which several electric generating facilities will be built to feed power through one or more transmission corridors to the neighboring population centers. Some jobs will be created for local workers, but in the main, the highly specialized well-drilling technicians and construction workers will come in from outside for temporary periods and then depart.

After demonstrating feasibility and safety in generating geothermal power, the installations will probably proliferate to a multitude of small plants, which may approach 1000- to 3000-MW capacity in aggregate. Some land will be diverted from agriculture, as the authors Pick and Butler point out. Some help will be developed for the energy shortfall, and some downward pressure will be applied to rising power costs.

Foreword

But again, as the authors describe, the major effects on land use, on life-style, on demographics, on the social structure and on the economy will be driven by the degree and amount of agribusiness and industrial development which associates itself with the geothermal resource. Several possible scenarios can be constructed, and some of the consequences can be predicted with a degree of sureness.

Edgar W. Butler and James B. Pick are experienced observers and qualified social scientists from the University of California, Riverside. They have, in this volume, attempted to analyze the data on population and economy which have been compiled about Imperial County. Fortunately, through the foresight of the county government and funding from Federal sources, voluminous base-line statistics have been compiled, and perceptions of people, and the plans of thought-leaders have been tabulated.

By use of regression analysis and other analytical tools, the authors are able to project the impact of the outcomes of several possible scenarios for development. These projections will be of great value in guiding policy makers in a period of change.

It is fortunate that a moment of history—the emergence of a stable, agricultural economy into a mixed economy with possible rapid growth—can be studied and analyzed and projected.

The effects of geothermal development may not be understood now. Many of them can be very beneficial; some can be adverse. But the effects can be managed if there is community understanding and effective leadership.

It is to be hoped that this volume will contribute to that understanding and will undergird effective leadership.

> Victor V. Veysey[*]
> Director
> Industrial Relations Center
> California Institute
> of Technology

[*]Former U.S. Congressman representing Imperial County.

Introduction

Energy development in a natural resource area such as the Imperial Valley proceeds by surges and ebbs like the tide. This results from the contribution of new technology and new market conditions which stimulate a wave of resource development. Many of the geothermal prospects that can be developed with the economic incentives currently available will be tested and brought to market, or as has been the case in the past in the Imperial Valley, set aside pending changed market conditions. The Imperial Valley has historically followed this pattern. In the 1920s, there was a period of exploration and test drilling in the area of the Salton Sea at Mullet Island. This resource proved to be unable to compete with surplus natural gas as an energy resource. It was followed in the early 1960s with another round of exploratory drilling which uncovered the hypersaline high-temperature brine underlying the Salton Sea geothermal area. Careful evaluation of this resource by several companies showed that it also was not competitive with natural gas as an energy resource and was not competitive as a potash resource because of the low cost of Canadian potash. In the early 1970s, another round of exploration and development started with drilling at Heber, North Brawley, East Mesa, and the Salton Sea. By the late 1970s, development expanded into the Westmoreland area and into the South Brawley area near the town of Imperial. Small 10-MW power plant facilities are being developed at East Mesa and North Brawley, with others planned for 1980 and 50-MW plants announced for East Mesa, Heber, and the Salton Sea areas. It appears as though the 1980s will be the decade of commercial development.

It is rare that major energy resource can be developed with intense scrutiny of environmental and sociological factors from the incipient stages of commercial development through to full development. The Imperial Valley may become one of these exceptional examples. The University of California at Riverside has attempted to place into the public domain a considerable body of information concerning both the scientific aspects of the resource and the sociological aspects of the early phases of development of geothermal energy. The UCR group combined effort in the sociological area is trying to reduce a large mass of data and a series of studies into a comprehensive book for the general reader with a strong technical background. As a former UCR faculty member, I can appreciate the incredible amount of work that has gone into preparing this book, and invite the reader to peruse it chapter by chapter, as the sections to a significant degree can stand independently. The depth and breadth of this effort is far greater than might appear on first reading, and as a reviewer of the various chapters, as well as someone who has spent nearly 30 years working in the Imperial Valley, I recommend this book to those who would like an insight into the impact of energy development in one of the more coherent communities in the western United States. What makes this study so important is the concentration in one isolated region of a remarkably large energy resource. It is, in essence, a natural laboratory for sociologists and economists to monitor to see how energy resource developers and an established community work together for better or for worse. It is the hope of all of us involved with this volume that it aid this effort in a constructive way.

Robert W. Rex
President
Republic Geothermal, Inc.

Preface

Part of this geothermal research grew out of the National Science Foundation (NSF) and Department of Energy's (DOE) Imperial County project.† The goals of that particular study were as follows:

1. To assess the extent of the resource as a means of determining the potential value of geothermal development versus other economic activity in Imperial County.
2. To determine alternative technologies and production costs for developing geothermal resources as a means of testing feasibility, cost, and value of geothermal resource.
3. To determine agricultural losses, as land is taken out of agricultural use and put into geothermal use, as a means of comparing the relative value of the two resources.
4. To determine the social costs (or human effects) for the purpose of abating the impact on people, employment, and social changes.
5. To determine environmental effects as a means of abating noise, subsidence, waste-water, air pollution, or other adverse ecological impacts.
6. To estimate the economic trade offs involving all the direct and indirect benefits and costs of geothermal development as a means of showing gains and losses to the various interest parties in Imperial County.
7. To assess the political interest groups and intergovernmental jurisdiction as a means of resolving conflicts among goals and arriving at the public interest.
8. To utilize the research for decisions as a management means of

†DOE was the Energy Research Development Agency (ERDA) at the time of this research. The NSE/DOE part of the research was carried out under grant No. AER 75-08793.

arriving at a workable geothermal plan (Internal Project Document, May 30, 1974).

The goals of this project arrived out of discussions by representatives of the Imperial County government and the University of California, Riverside, and California Institute of Technology. The County wanted to create an operable geothermal element to incorporate into its general plan. The general goal was to facilitate development of geothermal resources at feasible locations, mitigate undesirable environmental or social impacts of geothermal development, and provide for reasonable and effective regulation of the geothermal resource development. A second project was funded by DOE through the Lawrence Livermore Laboratory allowing a detailed study of leadership and geothermal development in Imperial County.

Appreciation is hereby expressed to all of the people who made this project possible; this includes people at ERDA/DOE, NSF, UCR, Cal Tech, Lawrence Livermore Lab, project investigators and their staffs, Tae Hwan Jung interviewers and their supervisors, and the citizens and leaders of Imperial County. Dr. Shawn Biehler, and Dr. Wilfred Elders of the Earth Science Department, University of California, Riverside, made valuable comments on earlier versions of this manuscript. Mr. Charles Hall, Lawrence Livermore Lab, was instrumental in the leadership study and made valuable contributions; Robert Lundy, Argonne National Lab, Dr. Robert Rex, President, Republic Geothermal, Inc., Victor Veysey, former Congressman from Imperial County and currently Director of the Industrial Relations Center, California Institute of Technology, and an anonymous reviewer, read the manuscript and helped us in many ways. Finally, Dr. Martin Pasqualetti, Department of Geography, Arizona State University, contributed to all phases of our research, and our relationship with him has added immensely to this research. All of these scientists, from a variety of disciplines, strengthened this research and were instrumental in shaping the final version of this book. Our sincere appreciation is hereby expressed.

University of California, Edgar W. Butler
   Riverside James B. Pick

Contents

Chapter 1. An Integrated Assessment Model for Public Policy
Alternatives, Priorities, and Outcomes in Geothermal
Environments 1
    Introduction 1
        Transferability of Research Results in Other KGRAs 2
        Assumptions 2
    An Orientation to Imperial County 3
    County Demographic and Employment Features 8
    The Research Model 14
    Functions of the Model 17
        Consolidation 17
        Systematization 18
        Identification 18
        Enhancement 18
        Mitigation 18
        Results 19
    Methodology 19
        Introduction 19
        Research Site Selection 20
        On-Site Research Data Collection 20
        Measurement of Key Factors 20
        Data Analytic Techniques 26
    Research Conclusions, Recommendations: Public Policy
        Alternatives, Priorities, and Outcomes 27
    Imperial County Research Prospective 30
    The Imperial County Plan 30
    Scope of Planning and of Energy Planning in Imperial
        County 31
    Acknowledgment and Comment 34

Chapter 2. Geothermal Energy in Imperial County, California 35
    Geothermal Resources 35

    Definitions 35
    Underlying Geological Theory of Geothermal Energy 35
    Types of Geothermal Resources 40
    Geothermal Exploration and Drilling 43
    Geothermal Power Plants 46
  Use of Geothermal Resources 58
  Geothermal Resources in Imperial County 60
    Origin of the Geothermal Resource 62
    Estimation of Geothermal Temperature and Reservoir
      Capacity 64
    Geothermal Area Physical Properties 66
    Effects of Fluid Withdrawal and Reinjection 67
    Life Expectancy of the Geothermal Resource 68
  Conclusion 74
  Acknowledgment 75

Chapter 3. Population–Economic Data Analyses Relative to Geothermal Fields, Imperial County, California   77

  Introduction 77
  Population Analysis of Imperial County 79
  Migration 86
  Energy Capacity and Consumption in Imperial County 88
  Population Composition and Change 89
  Population Economics 100
  Housing and Transportation 103
  Regional Socioeconomic Comparisons 107
  Migration and Income Characteristics 116
  Discriminant Analyses of Geothermal Areas 122
  Applications to Geothermal Energy Development 128
  Spanish-American Population of Imperial County 129
    Resident SAs 129
    Illegal Mexican Aliens 133
    The MA Population and Geothermal Energy
      Development 136
  Acknowledgment 137

Chapter 4. Regional Employment Implications for Geothermal Energy Development in Imperial County, California   139

  Introduction 139
  Regional Regression Analysis for Employment 146
  Analysis Result 151
  Implications of Geothermal Development on Regional
    Employment 161
  Acknowledgment 164

Chapter 5. Projected Population, Growth, and Displacement from Geothermal Development — 165

Introduction 165
Farm Labor Force Reduction Based on Land Area Analysis .. 166
Farm Labor Reduction 172
Historical Migration and Age Structure 176
Population Projections 177
    Series (I) and (II) 180
    Series (III) and (IV) 181
    Series (V) and (VI) 183
    Results 183
    Conclusions 190
Transferability of Methods and Results 194

Chapter 6. Public Opinion about Geothermal Development — 199

Introduction 199
Methodology 201
    Sample Design 201
    Questionnaire Design and Review 204
    Mail-Back Procedure 204
    Response Rates 205
    Cautions 205
Public Opinion about Geothermal Development in
    Imperial County: 1976 205
Understanding of Geothermal Development 210
    Background Characteristics 210
    Attitudes 211
Adequacy of Information about Geothermal Development ... 216
    Background Characteristics 216
    Attitudes 216
    Conclusions 218
Regulation of Geothermal Development 219
    Background Characteristics 219
    Attitudinal Statements 219
Foreseeing Problems Arising Out of Geothermal
    Development 221
    Background Characteristics 221
    Attitudes 222
Land Ownership and Opinion about Geothermal
    Development 226
    Land Ownership 226
    Landowner's Opinion 227
Rural Land Ownership 228
Public Opinion in the Heber Area 230

Conclusions and Recommendations 232
   Introduction 232
   Recommendations 232
   Transferability of Research 233
   Future Research 233
Acknowledgment 234

Chapter 7. Leadership, Community Decisions, and Geothermal Energy Development: Imperial County, California   235

Introduction 235
The Power Structure in Imperial County 236
   Methodology 236
   Influential People and the Power Structure in Imperial Valley 237
   Summary 239
Leadership Opinion and Reaction to Geothermal Development 240
A Comparison of Leadership Opinion and Public Opinion of Geothermal Resource Development 241
The Efffect of Geothermal Resource Development on the Power Structure in Imperial County 244
Conclusions 247
Acknowledgment 248

Chapter 8. Geothermal Development Update   249

Introduction 249
Beginnings of Geothermal Impact on County Population and Leadership 250
Electrical Use Developments 264
   East Mesa 264
   Heat Exchangers 266
   Pumps 266
   Cooling System 266
   Heber 267
   Brawley 268
   Salton Sea 270
Direct Use Developments 271
   City of El Centro Community Center 274
   Holly Sugar 279
   Valley Nitrogen 282
Regulatory Permitting Process in the County 287
Limiting Factors on Geothermal Development 289
   Transmission Lines 290
   Financing 290
   Political Factors 294

Direct Use Transmission Pipes 295
Cooling Water Availability 296
Brine Waste Disposal 303
Conclusions 303

Chapter 9. Research Conclusions and Policy Recommendations about Geothermal Development   305

Introduction 305
Major Research Conclusions 306
    Environmental Impact 306
    Population and Economics 313
    Policy Recommendations of the Population Analysis 318
    Public Opinion 319
    Research Conclusions on Public Opinion 322
    Policy Recommendations from the Public Opinion Survey   323
Recommendations 324
    Assumptions 324
    Environmental 327
    Socioeconomic 328
Implications of Research Oriented Planning to Imperial County and Other Counties 330
Acknowledgment 331

Appendix: Public Opinion Questionnaire 335
References 341
Subject Index 353
Author Index 359

1

An Integrated Assessment Model for Public Policy Alternatives, Priorities, and Outcomes in Geothermal Environments

INTRODUCTION

During the course of a research project on geothermal development in Imperial County, California, it became apparent that a broad perspective was essential if public policy alternatives, priorities, and possible outcomes were to be systematically delineated. Among the reasons for this belief are the following:

1. Geothermal energy sources could become important in filling the energy needs of the nation.
2. Research insights are necessary if rational long-term alternative energy policies are to be developed.
3. Overlap and duplication of research efforts often slow development and waste funds, especially when there is lack of communication and overlapping efforts.
4. Optimal mechanisms are needed to determine site locations vis-à-vis economic, social, and environmental costs.
5. Mechanisms are needed to expedite the decision-making process to speed geothermal resource development and to aid in formulating long-range national, state, and localized geothermal energy policies.
6. Communities need to prepare for changes prior to development of geothermal and other energy resources.
7. Information which can be applied to local geothermal planning

elements is necessary if local jurisdictions are to have any input and control over their resource development.

Transferability of Research Results in Other KGRAs

Since the research and planning model outlined in this chapter could incorporate geothermal resource areas in different stages of development, it allows improved transfer of research results from one site to another as known geothermal resource areas (KGRAs) develop. For example, the vapor-dominated area at the Geysers in California is in a commercial stage of development while the water-dominated area of Imperial County, California, is likely to be the next area of extensive geothermal development in the U.S. Other geothermal sites in California are undergoing preliminary development, while numerous sites in Utah, Nevada, and New Mexico will probably undergo development (Anderson, 1979; Rex, 1978)

Cutting across various resource types and environments are a variety of physical, socioeconomic, and political environments, as well as a host of local policies. Geoengineering research data are sometimes available for sites (see White and Williams, 1975; Muffler, 1979). Missing until now has been the integrated consideration of socioeconomic, environmental, population, and political spheres of data in research and planning models allowing the expedited transfer of research methods from one KGRA to another.

Assumptions

A number of premises or assumptions are necessary for a basic understanding of the research model we propose. Among them are the following:

1. The broad makeup of a geothermal productive area, including its social and demographic makeup, should be generally understood early in the development process.
2. The benefits of an energy resource and particular location should be understood and addressed early in development planning.
3. Social planning concerned with geothermal development should be undertaken as soon as possible prior to development.
4. Because research at the Geysers and in Imperial County, California represents a level of support not be generally repeated,

generalized knowledge gained from these sites must be made available for meaningful decision making on alternative policies, priorities, and potential outcomes of geothermal development for other KGRAs.
5. An example of commercial development should be encouraged even at some minor environmental risk. This can be done most effectively by having knowledge of a variety of resource areas and the environmental, social, and economic factors involved in each.
6. Generally, environmental restrictions can be placed on a sliding scale, becoming more stringent as technology improves.

Each of the above points needs to be considered if the decision-making and planning process of geothermal development is to be expedited as an alternative energy source. Before proceeding to geothermal development in Imperial County, a brief orientation to the county is necessary.

An Orientation to Imperial County

All active geothermal programs in Imperial County are concentrated in or immediately adjacent to irrigated fields in the Imperial Valley Trough. As shown in Figure 1–1, the Valley itself is part of the Salton Trough, a heavily faulted, low-lying structural depression about 200 km long and up to 130 km wide, which begins as a landward manifestation of the Gulf of California. The Salton Trough and the Gulf of California represent the boundary between the Pacific and North American crustal plates, a boundary extending north-westerly for over 900 km as the San Andreas Fault.

Figure 1–2 shows the Known Geothermal Resource Areas (KGRAs) and major geological faults in Imperial County. KGRAs are areas defined by the Federal Government as containing potentially exploitable geothermal resources. Activity along the San Andreas and other nearby faults such as the San Jacinto, Elsinore, and Imperial have produced more than 12 earthquakes of magnitude greater than 6.0 on the Richter scale during the present century. An earthquake on May 18, 1940 resulted in as much as 3.9 m of eastward lateral displacement along the Imperial Fault (Elders, 1975; Dibblee, 1954). Continued slippage along the Imperial Fault amounts to 3.9–7.8 cm per year (Sharp, 1972). In 1975, an earthquake swarm southeast of Brawley registered 36 earthquakes of a magnitude of 3.0 or greater (Elders, 1975).

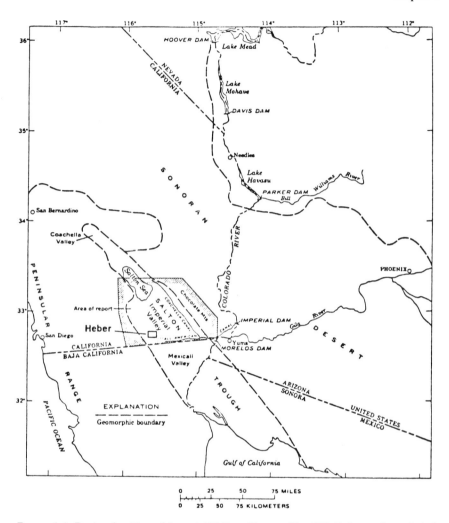

FIGURE 1-1. Regional setting of Imperial Valley. [Source: Vasel W. Roberts, *Geotechnical Aspects of Geothermal Power Generation at Imperial Valley, California*. Palo Alto, California: EPRI, a report prepared by Geonomics, Incorporated; Berkeley, California, EPRI ER-299, Project 580, Topical Report 1, (October 1976), p. 2.]

Coincident with the faulting activity, the Valley is subsiding and widening. The crustal spreading continuing throughout the Salton Trough is largely responsible for the concentration of geothermal resources found in the Imperial Valley (Elders, 1975).

Early agriculturalists and developers sought to camouflage the na-

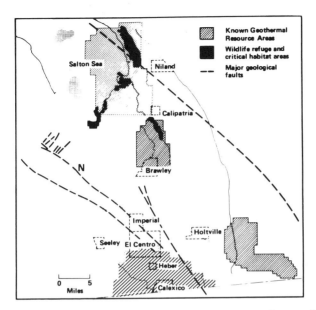

FIGURE 1-2. The Imperial Valley KGRAs which are considered to be suited for electric power production. Also shown are the wildlife refuge and critical habitat areas and the major geological faults in Imperial Valley.

ture of this region in south-central California by naming it Imperial Valley. Such ruses notwithstanding, it remains the hottest, driest, and most barren regions of North America. On the other hand, it is quiet, lightly settled, and almost always free of the photochemical smog that plagues metropolitan areas to the west. A continuing perception of inhospitableness by "outsiders" and a relatively new one of the cultural isolation by young locals have kept the population in a paradoxical "no-growth" posture while California as a whole has grown vigorously for decades. In 1975, the county had an estimated population of 84,100 residents, located mainly in the three major towns of Brawley, El Centro, and Calexico.

Even though the physical climate has remained the same, there have been great changes in Imperial County since 1900. The county was turned into an agriculturally productive region in 1905 through diversion of part of the flow of the Colorado River water onto the previous desert surface of Imperial Valley. Today, the result of that highly successful diversion is bountiful agricultural production. Imperial is currently California's second most productive agricultural county, yielding nearly $0.5 billion in farm products annually. Currently, the

Imperial Irrigation District states that the Imperial Valley is the "largest irrigated district in the world." Its 200,000 hectares of agriculture and 150 feed lots produce $0.5 billion of products yearly, supporting the majority of the Valley's 84,100 people, and have contributed to the rapid growth of neighboring Baja California, Mexico. The key to this miracle is over 2,800,000 acre ft of Colorado River water conveyed yearly through the All-American Canal and distributed by means of an elaborate irrigation and drainage system ending at the Salton Sea.

Imperial County, then, is one of the most productive agricultural regions in the world. It provides the United States with a large percentage of its winter vegetables. Other agricultural production includes sugar beets, cotton, and field grains. A third enterprise is devoted to raising forage and feed for local feedlots, as shown in Figure 1-3—a cattle feedlot near the Heber geothermal site, and the dairy industry serving Southern California. The main reason for the high productivity of the region is the 365-day growing season. The agricultural production capabilities of the Imperial Valley are essential for continued food production in the United States.

The major field crops grown in Imperial Valley average between 155,000 and 170,000 acres per year. Wheat is grown on between 100,000 and 150,000 acres, sugar beets on around 70,000 acres, and cotton on between 45,000 and 78,000 acres. A variety of other field crops are grown but with much smaller acreages. The major garden crops are melons with over 10,000 acres per year; about 45,000 acres of lettuce; 6,000 to 7,000 acres of carrots; and 5,000 to 7,000 acres of onions. The major permanent crop is asparagus with around 4,500 acres. Around 7,000 acres are used as duck ponds. The total acreage utilized hovers around the 600,000 level. The actual total area farmable is 476,000 acres, but land use may be accounted for several times because of the lengthy growing season.

In 1977, agricultural revenue totaled $437 million. This total was subdivided into field crops ($186 million), vegetable crops ($135 million), livestock and dairy ($102 million), seed and nursery ($9 million), and other ($5 million). The biggest county crop was lettuce ($76 million), followed by alfalfa ($64 million), and cotton ($61 million). Lofting (1977) estimated that agriculture accounted for 53.6% of gross county output (ignoring imports) in 1972.

The use of Colorado River water for irrigation has been facilitated by the concrete lining of the laterals furnishing water to the fields. By 1974, over 635 miles of the laterals had been concrete lined. A mechanism to flush salts from the fields was begun in 1929. Salt removal is accomplished by keeping the salts moving down through the soil by applying more water than the crop uses. The excess water carries away

FIGURE 1-3. Cattle feed lot near Heber, Imperial County, California.

the dissolved salts. In the Imperial Valley this process is facilitated by a drainage system of tiles to remove the leach water. Tile spacing in Imperial Valley ranges from 50 feet in silty clay soils to 300 feet in sandy soils. In 1974, over 20,000 miles of tiles had been installed in Imperial County. The leach water flows into the New and Alamo Rivers which empty into the Salton Seas (IID, Bulletin No. 1075).

The political environment of Imperial County differs from that of the state in degree, not in kind (Buck, 1977). The county registration and voting patterns increasingly reflect statewide trends. The county is undergoing slow political change with lower income and Chicano voters becoming more active. The principal political issues facing the county are unemployment, taxation, the 160-acre limitation, power rates, labor unrest, and just recently sewage disposal from Mexico and San Diego. A developing issue probably will be the location of electrical transmission lines.

Geothermal resources discovered in Imperial County are unique in that they lie mostly under highly cultivated lands. Most other geothermal resource areas have been discovered in undeveloped areas with less potential economic impact than is faced in Imperial County.

Culturally, Imperial County is devoted to agribusiness, and the desire of the elected officials and perhaps most of the population is to continue this agribusiness emphasis and at the same time enjoy the fruits of geothermal production. The official stated objective of elected county officials is that the two dissimilar activities coexist with minimal impact on agriculture. In Imperial County, seven principal efforts are underway to tap the geothermal resources underlying agricultural areas.

County Demographic and Employment Features

The present population composition of Imperial County has been determined by the unique history of the county. There are major demographic differences from other parts of California. Prior to 1900, Imperial County had only scattered settlers. The County consisted of a central desert surrounded on each side by rugged mesas and mountains. Diversion of a portion of Colorado River flow through the All-American Canal in 1901 created an irrigated central valley portion. This river diversion did not proceed without problems. In 1905 the Canal was broken apart and major flooding occurred in the valley—a disaster not ended until canal repair in 1907. This man-made scheme also vastly increased the potential for settlement, as an agricultural population was required to farm the land and the environment of the valley became less harsh for habitation. This river diversion must be regarded as the major event in the valley's history affecting its demography.

Another important influence on Imperial County's population history has been population growth trends in Mexico and border regulations. Table 1–1 shows the total population and decennial growth rates for Imperial County, California, and Mexico from 1910 to 1970. Imperial County grew rapidly until 1930 but has barely grown since that time. California, on the other hand, revealed steady and regular growth from 1910 to 1970, with decennial growth rates varying between 22% and 66%. California growth slowed down only in the 1930 Depression decade and during the 1960s. The failure of the county from 1930 to 1970 to match the state's growth rate is due to noncoastal, nonurban location, lack of educational and training opportunities, especially for the young, and lack of growth in the county job market. The latter reflects the dominance in the county economy of agriculture, which has nationally been characterized by employment decrease. In a pattern opposite to the county, Mexico grew slowly until 1940, but more rapidly since then. More rapid still has been the growth of Mexico's Northern Baja State, bordering the county. Northern Baja increased in population from 520,000 to 1,105,000 between 1960 and 1970 (Whetten, 1971). Mexicali, the largest city in Northern Baja, with an estimated 1975 population of half a million, is the sister city to the Imperial County town of Calexico (1970 population 12,078). This disproportionate Mexican growth, directly across the border, is an important factor in present and prospective county geothermal development. It will be alluded to often in later chapters, such as in sections of Chapter 8 on water supply and on transmission corridor planning.

TABLE 1–1. TOTAL POPULATION AND DECENNIAL GROWTH RATES FOR
IMPERIAL COUNTY, CALIFORNIA, AND MEXICO, 1910–1970[a]

	Imperial County		California		Mexico	
Year	Total population	Decennial growth rate (%)	Total population	Decennial growth rate (%)	Total population	Decennial growth rate (%)
1910	13,591	—	2,377,549	—	15,160,369	—
1920	43,453	220	3,426,861	44	14,334,780	−5
1930	60,903	40	5,677,251	66	16,552,722	15
1940	59,740	−2	6,907,387	22	19,653,552	19
1950	62,975	3	10,586,223	53	25,791,017	31
1960	72,105	18	15,720,860	49	34,923,129	35
1970	74,492	3	19,953,134	27	48,313,438	38

[a]Source: U.S. Bureau of the Census, 1910–1970.

Besides its population, another factor influencing interaction with Mexico is fluctuating international border regulations. This status historically has been characterized by major and often erratic swings. For instance, in the early 1900s up to 1930, legal immigration of Mexicans into the United States was immense and unrestricted. From the late 1950s to the mid-1960s, legal immigration from Mexico was very low, but Mexican labor nevertheless came into the U.S. under work contracts. Recently, legal immigration northwards has remained very low, while illegal migration has been at an all-time high. Future border regulatory changes undoubtedly will influence Imperial County's labor market and thus geothermal development.

The steady levels of Imperial County's population since the 1930s indicate a very large and persistent out-migration. Moreover, age structure analysis reveals a concentration of this out-migration in young age categories. Estimates of large past out-migration are based on the basic formula for population change:

$$\text{population change} = B - D \pm M$$

where

B = births in a time interval
D = deaths in a time interval
M = net migrants *into* an area $(M+)$ or *out of* an area $(M-)$ in a time interval

county population change has been very slight since the 1930s, and since births have greatly exceeded deaths because of high county fertility, computations reveal that net migration has been large and negative. In fact, it has averaged −0.7% per year since the 1930s. The youth of out-migrants is revealed by net migration rates for female age categories from 1960 to 1970. (Rates for female categories are considered to be more accurate than for male.) The rates were −10.3% for 0 to 29 year olds, +0.4% for the 30 to 64 year olds, and +9.2% for persons 65 and older. Young people have migrated out of the county to seek educational and training opportunities unavailable locally.

The county population by age and sex structure for 1970 is shown in Figure 1–4, a population pyramid. The vertical (ordinate) axis represents age groups in the beginning with the lowest (0 to 4) ages at the bottom and proceeding upwards to the oldest (85 years and older) at the top. Males are shown to the left of the center line and females to the right. A noticeable concentration of population for both males and females appears in the age categories 5 to 19. These categories comprised 35.1% of the 1970 county population, versus 29% for the U.S. 1970 population. This population is now (1979) 14 to 28 and has, undoubtedly, contributed many out-migrants during the 1970s. Another aspect of the age structure is a 15% excess of females over males in the young adult category 20 to 39. This excess contrasts with a 6% excess of males in this category for the U.S. Such a difference is the result of selectively higher net out-migration for males versus females. Since county males (at least in the 1960s) were more oriented towards occupational advancement than females, they would be more likely to move away to areas of greater educational job opportunity.

By ethnic composition, the county had a 45% Spanish-American component in 1970, a remarkable high percentage compared to only 14% for California. Such a large ethnic population has resulted in ethnic differences in many social and economic characteristics. This Spanish-speaking population group is concentrated geographically in the county. As seen in Figure 1–5, the greatest concentration is in the town of Calexico (population 10,625), with other concentrations in the rural areas around the town of Brawley and in the southwestern agricultural areas. Calexico, which was 91% Spanish speaking in 1970, contained 28.3% of the county's entire Mexican-American population.

This large and concentrated population shows differences from the anglo population in many population, social, and economic characteristics. Fertility is much higher for Spanish-Americans. In 1970, the average ever-married Spanish-American woman age 35 to 39 had borne 4.9 children, versus 3.3 for an anglo woman. As might be expected from so many births, the Spanish-American age structure is much younger than the anglo structure. An educational gap is re-

An Integrated Assessment Model

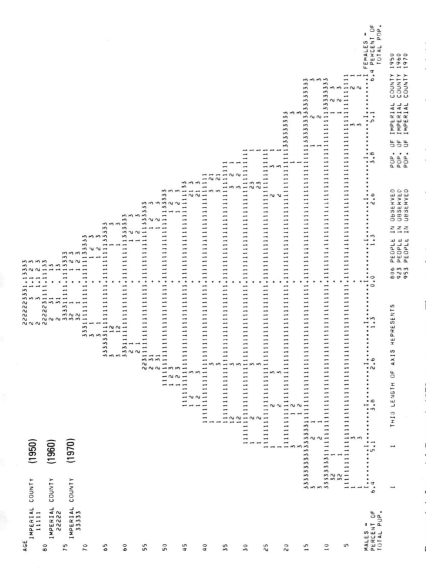

FIGURE 1-4. Imperial County's 1970 age structure. The population pyramid reveals large numbers of children and adolescents. In 1980, members of these groups will be ages 15–29, and many will have out-migrated from the county based on past precedent.

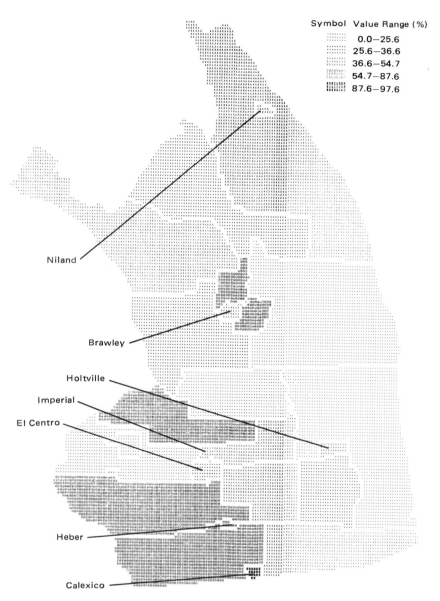

FIGURE 1-5. Percent of Spanish-speaking population: Imperial County, 1970. (Source: ED Computer tapes, U.S. Bureau of the Census.)

TABLE 1-2. INDUSTRIAL CATEGORIES IN IMPERIAL COUNTY, 1950-1977

Industrial category	1950[a] (percent)		1960[a] (percent)		1970[a] (percent)		1976-77[b]	
	Male	Female	Male	Female	Male	Female	Both sexes (percent)	Both sexes (number)
Agriculture	43.4	8.6	48.4	6.3	25.1	8.8	40.2	15,475
Construction–Mining	6.6	0.6	4.3	0.4	7.5	1.0	2.1	825
Manufacturing	7.1	2.0	6.6	1.9	8.5	3.2	5.4	2,075
Transportation–Utilities	9.7	7.8	8.9	5.0	11.8	4.9	3.2	1,250
Wholesale–Retail	17.8	33.0	14.7	28.3	22.4	28.7	18.2	7,025
Government	4.9	13.6	6.2	16.2	12.3	18.1	21.9	8,450
Other	10.5	34.4	10.9	41.9	12.4	35.3	9.0	3,425
Total number of workers	18,599	4,518	21,613	6,414	15,397	8,082	38,525	

[a] Data from U.S. Bureau of the Census, 1950-1970.
[b] Data from Employment Development Department (EDD), State of California, 1978.

vealed by achievement of high school graduation for 55% of anglos versus 23% of Spanish-Americans. Similarly, average 1970 family income was $10,842 for anglos versus $7,613 for Spanish-Americans. Such differences are similar for all socioeconomic characteristics; these data are analyzed in detail in Chapter 3.

The county's employment distribution from 1950 to 1976–1977 is shown in Table 1–2. It is not surprising that agriculture accounts for the largest share of employment. The figures for 1976–77 from the State's Employment Development Department (EDD) are more comprehensive and hence more accurate than the earlier U.S. Census figures, especially for agriculture, for which the EDD includes border commuter farm workers. The large component of 15,500 agricultural workers consists mostly of very low-skilled and low-wage farm laborers.

Other large employment sectors are government, wholesale–retail trade, and manufacturing. The service sector, on the other hand, is significantly smaller than for the state. The 27-year growth in the government sector to comprise 36.6% of nonagricultural employment in 1976–1977 is remarkable, especially since government sectors only account for 4.6% of nonagricultural gross county output. The manufacturing segment is somewhat smaller than that for California, eliminat-

TABLE 1-3. EMPLOYERS IN IMPERIAL COUNTY, 1976[a]

Company	Products	Number of employees
Holly Sugar Company	Sugar, molasses, beet pulp	400
U.S. Gypsum	Gypsum	280
Valley Nitrogen	Fertilizers	161
Dessert Seed Company	Seed processing	150
Ryerson Concrete	Concrete products	105
Douthitt Steel and Supply Company	Industrial supplies, oxygen, medical gasses	100
Imperial Valley Press	Daily newspaper	86
El Centro Garment Company	Garments	75
Anthony Williams	Women's wear	47
		1404

[a]Source: El Centro Chamber of Commerce, 1976.

ing agriculture in the comparison. As seen in Table 1-3, it is almost exclusively concentrated in nine large companies.

In spite of the county's wealth of agricultural output, income, poverty, and unemployment figures, all indicate an average standard of living below that of the state. For instance, in 1970, 16.1% of the county families were below the poverty level, versus 11.9% for California. County male unemployment was 7.1% versus 4.2% for the State. This rate has risen further, as indicated by 1977 EDD figures for the county of 21.2%

Other demographic aspects of Imperial County are examined in more detail in Chapters 3 and 4. Chapter 5 presents projected population growth and displacement from geothermal development.

THE RESEARCH MODEL

The goal of the research model used in this research effort was to expedite geothermal development. To achieve this goal, the model systematically examined the following: (1) *extralocal conditions*, such as national and state economic status, federal and state political structures, and federal and state policies; (2) *site-specific characteristics*, especially various elements of the physical environment, population and social characteristics, and economic and technological bases; (3) *local political processes*, including voting, political organizations, citizen participation, the articulation of the demand for services, and the decision-making process; (4) *local political system* in regard to governmental forms, legal boundaries of jurisdiction and representation, organizational characteristics, and leadership systems, all of which lead to, (5) *localized public alternatives, priorities, and outcomes* including, among others, plant site location, esthetics, water use, local versus distant use, transmission, and electric versus nonelectric possibilities. These dimensions of the model are shown in Table 1–4.

The systematic utilization of these various research and planning elements allows (1) the use of generally available information in decision making; (2) the establishment of priorities; and (3) the predetermination of many outcomes prior to geothermal development.

The basic research and planning model shown in Table 1–4 outlines a variety of factors related to localized public policy alternatives, priorities, and outcomes. Among the outcomes, at least the following must be considered:

(1) Decisions must be made on how to protect the quality of life for citizens, including their health and property. This includes various kinds of safety measures and involves environmental monitoring and mitigation of environmental impacts on air, water, and in some instances, subsidence and seismicity.

(2) Consideration must be given to the question of how to optimize assets and resources of the community located near the KGRA. Among the factors that need to be considered are employment, tax revenues—both generation and expenditures, and impact on historical, biological, and local archeological resources.

(3) Decisions have to made about life-styles that exist in the community now and may subsequently exist as a result of geothermal or other resource development. Among the considerations are those involving the possibility of rapid growth and development of "boomtowns," economic and/or tax burdens, and various kinds of aesthetic considerations such as transmission lines, pipe lines, electrical girds, etc.

(4) There must be systematic evaluation of plant site location and plant sizes, especially in regard to noise, odor, and wind factors that might bring unwarranted deterioration of the local environment. Included would be how to minimize or maximize preferred impact on agriculture, surface land, towns and cities, and how local control might be maintained with minimal external influence.

(5) After geothermal development there must be a public policy about alternatives and priorities related to how the burden and benefits are going to be distributed. Potentially, geothermal development could cost a local community more than it brings in revenue. Who is

TABLE 1-4. A Model for an Integrated Assessment of Geothermal Environments and Public Policy Alternatives, Priorities, and Outcomes[a]

(1) Extralocal conditions	(2) Geothermal site-specific characteristics	(3) Local political systems	(4) Local political process and outcomes	(5)[b] Localized public policy alternatives, priorities, and outcomes

A. National and state economics
B. Federal and state political structures
C. Federal and state policies
  1. Land acquisition
  2. Exploration
  3. Exploratory drilling
  4. Developmental drilling
  5. Production
  6. Utilization
D. Cost of energy from alternative sources
  1. U.S. Government policy
  2. Foreign
  3. State policy
  4. Environment tradeoffs
	A. Physical environment
	  1. Elevation
	  2. Topography
	  3. Biology
	  4. Etc.
	B. Population and social characteristics
	  1. Size
	  2. Ethnicity
	  3. Age structure
	  4. Etc.
	C. Economic and technological base
		A. Government forms
		B. Legal boundaries of jurisdiction and representation
		C. Leadership system
		D. Organizational characteristics
		E. Regulations
		  1. Land acquisition
		  2. Exploration
		  3. Exploratory drilling
		  4. Developmental drilling
		  5. Production
		  6. Utilization
			A. Voting
			B. Political parties
			C. Political organization
			D. Citizen participation and public opinion
			E. Articulation of demand for services
			F. Decision-making process
				A. Physical environment
				B. Population and socioeconomic environment
				C. Political, government, and regulations

[a] Derived from Paulson, Butler, and Pope (1969); Butler (1976); and Butler (1977).
[b] See text for specific outcome areas under each of the below.

going to share this burden? Alternately, extensive new tax revenues may be engendered for the community. How these resources are to be utilized, distributed, etc., may become an important part of policy alternatives and priorities.

(6) Among various policy alternatives and priorities are those related to specific KGRAs in regard to the various kinds of technology that are going to be used and what particular elements are to be involved in specific plant and site development. Among these are the consideration of how cooling towers are to be designed, to what extent mufflers will be utilized, and where hot water transmission pipes and electrical transmission lines will be located, among others.

Functions of the Model

Resource development is often a haphazard, protracted affair. The function of the integrated assessment model is to streamline the process of geothermal energy development by the use of six steps: (1) consolidation, (2) systematization, (3) identification, (4) enhancement, (5) mitigation, and (6) results.

Consolidation

Resource development is often encumbered by poor communication of ideas and data. Part of the problem is found in the dispersed nature of the literature in public libraries, university special collections, and data located in companies' files, computer tapes, archives, and obscure government research reports. Were this information available quickly, in a consolidated form for each of a number of key geothermal sites, developmental progress could be speeded up. Thus, one function of the research model is to consolidate data in the model for the variety of KGRAs.

One attempt at such consolidation was a data analysis of 37 sites in the western U.S. performed by MITRE Corporation under DOE sponsorship. Results were presented in a large three-volume report (Leigh *et al.*, 1978; Trehan *et al.*, 1978; Williams *et al.*, 1978). This consolidation, however, was focused almost entirely on the geotechnical resource, plant financing and construction, and regulatory hurdles. The only items from the present model (Table 1–4) which were more than cursorily covered were items 1D (cost of energy from alternative sources), 3E (regulations), and 5A (physical environment).

Systematization

Once data are accumulated, they often lie idle for want of organization and systematization. The logical and useful arrangement of data is as important to its expeditious use as its collection. A vitally important function of our research model is to organize and systematize consolidated data in such a manner as to ready it for useful application at geothermal sites as well as areas with other energy resources.

Identification

The model outlines those aspects of the resource in a variety of specific sites which distinguish it from the resource in other locations. Each type of resource, be it water-dominated or some other type, has its own advantages and disadvantages, and much the same can be said for individual KGRA sites. Imperial County, California, for example, has the advantage of nearby markets but the disadvantage of an irrigation and drainage network susceptible to the effects of slight subsidence, whereas The Geysers does not have agricultural crop or subsidence concerns but its chemical composition produces foul smells. Relative advantages and disadvantages of each specific chosen site for geothermal development have to be examined in terms of how they affect geothermal development and the local community.

Enhancement

Lack of identification and/or complete understanding of individual sites for potential geothermal development may inhibit the full utilization of some of the advantages of a particular site. A major function of the research model is to enhance or amplify those aspects contributing to a greater, quicker, and more efficient use of the resource, but which will not degrade the environment. In Imperial County, originally, little attention was given to the coincidence of agriculture and the geothermal resource. Because of this oversight, the full possibility of nonelectric uses escaped examination until only recently.

Mitigation

At every site of possible geothermal development, certain factors have a tendency to inhibit successful, efficient resource utilization. By early identification of these impediments, processes may be instituted to deal earlier with the more significant ones. An evaluation of the vacation-home land use downwind of The Geysers geothermal plant in Sonoma County, California might have alerted developers to the possibility of citizen complaints of sulfur odors which effectively halted construction programs for several years. Advance recognition of the possible significance of this problem might have resulted in an early direction of attention to hydrogen sulfide abatement devices or at least to a public education program aimed at the spread of accurate information on odor and harmful effects of hydrogen sulfide. Transmission corridor planning in Imperial County, as a further example, should have been an early focus, but it was left until last even though it may end up being the most critical factor in the development of large-scale geothermal energy. In an early phase of development, limiting factor analysis appears essential.

Results

The five steps or functions discussed above have one goal: to facilitate and speed the broad scale, efficient use of one type of energy resource which is apparently rich in promise but, so far, poor in fruition. Conceptual conversion to the availability and use of new energy resources is probably as significant as technical conversion, and it is the function of the research model to speed both.

METHODOLOGY

Introduction

The research model specified in Table 1–4 requires certain kinds of information. Below we elaborate upon measurement, data acquisition, and data analytic techniques required by the research model. Each particular dimension that will be necessary to carry out the research project specified in Table 1–4 is explored. The data necessary for the model are substantially available in a variety of currently unsystematized and discrete sources. The model brings together various data in order to carry out more systematic analyses which should lead to various kinds of localized public policy alternatives, different kinds of priorities, and outcomes based upon particular elements in regard to extralocal conditions, KRGA site specific characteristics, local political processes, and local political systems. In other words, the model utilizes various data sets systematically in order to identify a variety of public policy alternatives and priorities that can be applied to a variety of geothermal sites, depending upon the various kinds of environmental and social conditions as they exist differentially in selected local geothermal sites.

Research Site Selection

The model applies to all KGRAs, as well as to areas undergoing other energy resource development.

On-Site Research Data Collection

Most of the data for the model to evaluate any given site are already available, but in an unconsolidated and unsystematized form except for the MITRE reports. In addition, on-site visits provide substantial insight into the local political system and processes, public opinion, and leadership views regarding geothermal development. On-site visits should take place at the beginning of a project as well as during later phases to check externally gathered data as well as verify locally developed policy alternatives, priorities, and outcomes.

Measurement of Key Factors

The model presented in Table 1-4 requires data from a variety of sources. These bits of information are at various levels of measurement. Our procedure was to obtain data wherever they were available and to consolidate them into a form amenable to systematic computer analysis. Measurement of each dimension specified in the subsequent Policy Alternatives section is made possible from available sources and by on-site research as specified above. Specific sources and measures are delineated below in a fuller exploration of the key dimensions shown in Table 1-4.

Extralocal Conditions. There are a variety of extralocal conditions affecting geothermal development. Among them are at least the following:

A. National and state economics, that is the state of the national economy, e.g., inflation and/or recession. In addition, the national energy policy and its application influences the extent and pace of development by enabling legislation and appropriating funds for research and development. The Federal impact upon interest rates, loan guarantees, etc., also influences geothermal development. Similarly, at

An Integrated Assessment Model

the state level, laws giving state tax breaks, etc., have an impact upon development. Both Federal and state taxation, securities regulations, and utility regulations influence geothermal development.

B. Federal and state political structures' and climates' impact on geothermal development. For example, if a shortage of oil occurs again the U.S., undoubtedly the Federal and state governments would be able to move quickly in developing alternative energy sources with public support.

C. The following specific Federal and state policies' impact on geothermal development:

Federal
1. Land acquisition:
   Bureau of Land Management
   Environmental Impact Reports Approval
2. Exploration:
   Federal Regulations, generic
   Bureau of Land Management Exploration Permit
   U.S. Geological Survey Conservation Division
3. Exploration drilling:
   Federal Regulations, generic
   U.S. Geological Survey Conservation Division
   Bureau of Land Management Exploration Permit
4. Development drilling:
   Federal Regulations, generic
   U.S. Geological Survey Conservation Division
   Bureau of Land Management Exploration Permit
   Environmental Protection Agency
5. Production:
   Environmental Protection Agency
   Bureau of Land Management Exploration Permit
   Federal Trade Commission
6. Utilization:
   Federal Regulations, generic

State (e.g., California)
1. Land acquisition
   State Land Commission Permit to Explore
   Environmental Impact Reports Approval
2. Exploration:
   State Regulations, generic
   State Land Commission Geologic Permit on State Lands
   Department of Oil and Gas Regulations
   Drilling Regulations

3. Exploration drilling:
     State Regulations, generic
     Drilling Regulations
     State Energy and Conservation Agencies
     State Land Commission Permit to Explore
  4. Development drilling:
     State Regulations, generic
     Drilling Regulations
     Regional Water Quality-Control Board
     State Land Commission Permit to Explore
  5. Production:
     Public Utilities Commission
     Air Resources Board
     Energy Commission
     Regional Water Quality-Control Board
  6. Utilization:
     State Regulations, generic
     Marketing, generic
     Public Utilities Commission

D. The development of energy from alternative sources, as influenced by foreign, U.S., and state policies and environmental trade offs are of varying degrees of importance at different times and places.

Geothermal Site–Specific Characteristics. Under site-specific characteristics, the primary elements examined are (1) the physical environment, (2) population and social characteristics, (3) the economic and technological base.

(1) Physical environment. In an examination of the physical environment, at least the following sources of information can be utilized: U.S.G.S. Circular 790 (Muffler, 1979) and its predecessor 726 (White and Williams, 1975), which represent the only systematic efforts so far to evaluate the nation's geothermal resources by a common method. Their comparative value is thus substantial and their data are useful in initial phases of localized studies.

Data include location, subsurface temperature, volume, heat content, recovery factor, and electrical potential data (see Tables 5 and 6 of U.S.G.S. Circular 790). Also drilling and thermal data are given for potential low-temperature areas (Table 12), and temperature, geothermal gradient, and salinity data are presented for thermal springs in the central and eastern U.S. (Table 13).

Another large repository of available data is the U.S. Weather Service computer bank at the National Climatic Center located in Asheville, North Carolina. The climatological data available from the National Climatic Center includes average temperatures and departures from normal, and total evaporation and wind movement (1977). The computerized data of the Geothermal Information System (GRID) being established at the Lawrence Berkeley Laboratory, though incomplete, are useful in their present state and additions are being made to them on a continuing basis.

Other publicly available information to be used includes material in the scientific literature, progress and final reports of federally funded research, as well as projects under the auspices of local, regional, and state agencies such as the California State Energy Conservation and Development Commission.

Existing remotely sensed imagery also are available for many California sites: space, U-2, and low level are available.† In addition, low-level, hand-held photography is available for many areas. Information also can be provided by local university communities in each geothermal area, on-going projects within development companies, on-site visits, and interviews with officials and local citizenry.

The physical environment can be divided into three major categories for study: subsurface, surface, and above-surface. Geology is the most important aspect in initial development and continuing phases of geothermal consideration; without the resource (and this connotes aspects which make it exploitable, such as salinity of the conducting fluid and permeability and porosity of the host rock) everything else is meaningless. Subsurface evaluation, thus, is the first made, but needs to be updated as data become available. Data from U.S.G.S. Circular 790 already referred to, and data from development companies should be utilized when possible.

The ground surface needs to be investigated with regard to those aspects which will help or hinder exploration and exploitation, including the physical, cultural, and economic landscape. Elevation, topography, and biological and water resources can make geothermal development impossible or encourage it. Accessibility, the value or rarity of existing vegetation and wildlife, and certainly the existence of ample water supplies can be critical issues. In Imperial County, for example, the level terrain facilitates the movement of workers and equipment in contrast to the hilly terrain at The Geysers. In Imperial County, damage to crops is of concern because of their great value. In Imperial County, one of the most complex commodities is water.

---

†As an example, three different groups of scientists have made photographic flights over Imperial County in recent years, each apparently unaware of the others.

Our research model requires an understanding of the cultural nature of the landscape as well as its more virgin condition. How people changed the land and its sensitivities can either impede or encourage development. In Imperial County, the cultural landscape has presented many problems; fears of subsidence and crop damage from hydrogen sulfide would not exist without the cultural landscape. The existence of agriculture also is of extreme importance because of the obvious political organization and economic investment which helped create it and continues to profit from it.

An understanding of the economic interrelationships can mean the difference between success and failure in geothermal development. Distance to markets, labor supplies, mass transportation networks, support facilities, and their interrelationships with the regional resource base can be deciding factors in the economic feasibility of geothermal development. In Imperial County, markets for power are nearby, there is a large, local labor pool, there are limited amenities to attract workers but commuting is possible, and there are few other presently exploitable resources. All will have an influence on the success of geothermal development there.

The climate (and possibly the microclimate) in the region of each geothermal site has a bearing on the availability of water for cooling (or reinjection), the type of cooling tower allowed, disposal of wastes, construction times, worker comfort, and possible nonelectrical applications.

All these features are of significance in the future use of an area—for example, in terms of its value for recreation, vistas, settlement, commercial, and industrial development. In the consideration of the possible on-site vs. exported uses of the resource, such factors are critical.

(2) Population and social characteristics. Population and social characteristics can be gleaned primarily from *Population and Housing Census data,* as well as from any up-to-date local censuses that have been completed in particular geothermal locales, e.g., in Imperial County in 1975. These general population and housing reports give extensive information about the population, e.g., 1970 and 1980, and for previous decades. Thus, it is possible to examine trends as well as to make subsequent determination of whether or not geothermal development has altered the population characteristics of the particular site area. In addition, other resources are available from the U.S. government such as the computerized *City and County Data Book* which includes a variety of information in regard to public policy alternative, priorities, and expenditures in localized areas. These data are available

in computer tape format from the User's Division of the U.S. Census Bureau. Also available are the following kinds of data:

1. U.S. congressional publications.
2. Data from the Immigration and Naturalization Service (for counties on or near the Mexican border).
3. State vital statistics reports and publications.
4. State population projections.
5. State economic reports, such as the Employment Development Department publications in California.
6. Data from local planning departments (these data must be evaluated for accuracy).
7. Data contained in research literature on the locality. These data might be from special surveys.

Economic and technological base. Economic and technological base information is available from OBERS Publications as well as from the *U.S. Census on Manufacturing* and various other sources such as Dun and Bradstreet publications.

Local Political Systems. Elements of the local political system can be obtained from the *Municipal Year Book*, which specifies local government forms, expenditures, etc. The extent of overlapping legal jurisdictions in representation can also be obtained from various state and local publications giving insight into the intergovernmental complexity that exists in local geothermal development site areas. It is expected that there will be substantial variation by states and by local jurisdiction in this regard.

Local leadership systems can be determined by on-site visits and interviews with a variety of people living in the area as well as by utilizing the U.S. Census Bureau data to obtain data on organizational and leadership systems.

Finally, organizational characteristics of the localized area can be determined from local and state publications, interviews with local politicians, and others.

Local Political Processes. A variety of data are currently available in regards to local political processes. Among these data are voting behavior available by local counties, cities, and townships and within them by precincts. In addition, the *City and County Data Book* includes information about the predominant political party in various segments, including towns, cities, urbanized areas, and SMSAs. In order to determine the actual political organization of the area, field work and local contacts and publications can be consulted.

The extent of citizen participation also can be determined through local contacts, publications, and other unpublished sources. The articulation of demand for services can be, to some degree or other, ascertained through data included in the *City and County Data Book*: this includes expenditures for a variety of services including police, fire, education, health, etc., and gives a good indication how current tax revenues and other resources are distributed within a local jurisdiction.

Finally, the decision-making process and leadership structure can be determined through local contacts, organized analyses of available census and housing data, and through utilization of various published sources, especially local newspapers.

Localized Public Policy Alternatives, Priorities, and Outcomes. The generation of extralocal conditions, factors, geothermal site-specific characteristics, and evaluations of local political process and local political systems are all necessary in order to suggest localized public alternatives and priorities, and to generate expected outcomes of energy resource development. The original reason for this particular research was, of course, to generate differential public policy alternatives, priorities, and expected outcomes for Imperial County and the localized geothermal sites located in the county. Subsequent research results, using our model, will give other local public policy and public decision makers a more rational, knowledgeable, and reasonable background for making decisions, establishing certain kinds of priorities, and maximizing the kinds of outcomes that they would like to have as a result of geothermal energy resource development in their local area.

Data Analytic Techniques

The methods used for analyzing the above data can be descriptive and/or highly statistical. We present data in a tabular and graphical form for purposes of analysis of individual geothermal sites and comparisons among sites. For mapping of geoengineering and environmental data, hand and computer mapping techniques like Automap II Computer Graphics Program (ESRI, 1975) were used. Population age structures, for the last several censuses, displayed by PYRAMID (Pick, 1974), a line-printer computer graphics routine, are valuable and can be carried out for any local area. For population data, tables derived from U.S. Census data can be used to show important features and trends. Statistical analyses we have presented for Imperial County can be carried out for all localized areas. Economic–technological data are available from standard sources (e.g., *U.S. Census of Population*, *U.S. Census of Manufacturing*). In addition, the latest Dun and Bradstreet business tapes can be accessed in order to determine the economic and labor force structure of the private sector in geothermal site areas. Aggregated tables can be created for purposes of comparison.

A second category of methodology that is used is multivariate analysis. Data used for multivariate tests are mostly population and economic data; geoengineering and environmental information for the geothermal sites also can be used in the multivariate analysis. Correlation studies can be done on population and socioeconomic enumeration district data to determine how closely linked variables are within each site.

Discriminant function analysis is useful in determining key variables which distinguish geothermal areas from nongeothermal areas within site counties or elsewhere. For pooled data from a variety of sites, both of the above techniques can be used to determine variable linkages for the pooled areas, and distinguish features between geothermal and nongeothermal areas.

As the variables for individual and combined sites are better understood, regression analyses can be performed in order to examine the multiple determinants of key variables. Dummy variables can be used in the regressions to incorporate qualitative community and political data.

In order to assist in formulating generalized patterns of possible cause and effect relationships among the many socioeconomic variables, path analysis can be accomplished for individual KGRAs and for consolidated geothermal regions. Results of the path analysis, prior descriptive and multivariate studies, and qualitative conclusions from site visits and available literature can be used to formulate an integrated summary of the research—a policy flow table. The policy flow table is a tool useful in developing public policy alternatives and establishing priorities in relationship to specific outcomes.

Research Conclusions and Recommendations: Public Policy Alternatives, Priorities, and Outcomes

There is a need to develop models of public policy alternatives, priorities, and outcomes for various geothermal resource areas so that research results can be transferred from one geothermal area to another, and from one energy resource type to another.

Our research model facilitates geothermal development and assists in cost-effective development. Thus, the model will assist in starting new developments. For those political jurisdictions that feel it necessary to conduct their own research, our model serves as a prototype for local research efforts, and the data we have developed can be generated for all such sites.

The research model is transferable both in substantive and methodological respects. It alerts those unaware of the potential issues involved generally in geothermal development and those that are relevant only to certain sites. Our model can be used by local jurisdictions to develop plans in a systematic manner.

In basic terms, the model's conclusions and recommendations are presented in four forms:

1. Description of the environmental, political, and socioeconomic environments that exist in given KGRAs.
2. Enumeration of the possible benefits obtainable from development of the resource.
3. Delineation of the features of both the area and resource that might impede its development. These possible impediments can be described in environmental, regulatory, and social and economic terms.
4. Development of a policy flow table describing policy alternatives, priorities, and outcomes that are possible for each KGRA.

Below, in a general ordering or grouping by issue area, are most of the major elements which public policy alternatives, priorities, and outcomes must consider. All of them are of primary importance in the early developmental phases. Some are general to all geothermal sites and others are site specific.

Physical environment

1. Site location (e.g., size, slant vs. straight drilling)
2. Land use (e.g., agricultural, urban, etc.)

3. Subsidence
4. Seismicity

5. Water supply—quantity and quality
6. Water pollution
7. Water waste disposal—blowdown, reinjection, etc.
8. Reinjection

9. Air pollution—odor, chemical composition
10. Cooling towers—climatic

11. Site aesthetics—pipes
12. Power corridors—transmission lines, location
13. Noise pollution

14. Biological

15. Recreation
16. Cultural—archaeological, historical

Population and socioeconomic environment
1. Regulatory processes—federal, state, regional, and local
2. Extent of local vs. external control, pricing, gridding
3. Industrialization
4. Economic and technological base
5. Employment
6. Labor force composition
7. Secondary socioeconomic impacts
8. Institutional impact
9. Population—composition
   ethnicity/race
   age
   migration pattern
   education and training level
10. Final demand (taxes)
11. Disposal of tax revenues
12. Social services expenditure impacts

Political, government, and regulations
1. Land ownership, type
2. Leasing arrangements
3. Tax—policies
   use
   dislocation
   subsidy
   resource
4. Permitting fees
5. Leadership participation and opinion
6. Public participation and opinion
7. Government form
8. Voting
9. Political organizations
10. Mitigation measures
11. Regulatory agencies
12. Electricity production
13. Local vs. Nonlocal use of electricity
14. Nonelectrical (local-use)
15. Process heat production
16. Mineral products
17. Water production

IMPERIAL COUNTY RESEARCH PROSPECTIVE

This book does not report on all phases of the NSF/DOE and DOE research conducted in Imperial County. Rather, we focus on how geothermal development may affect the population composition, including some economic facets, employment implications, and possible growth and development patterns. Further, we systematically explore citizen or public opinion about various facets of geothermal development, about how the "leaders" feel about it, and compare public opinion with that of the leaders. In addition, we explore what impact geothermal development may have on the citizens and leaders of Imperial County.

Since this research was conducted during the exploration stages and during the development of the county's *General Plan*, we also have updated necessary factors concerning geothermal development in the county. Our goal was to help citizens and leaders of Imperial County more effectively utilize the geothermal resource for their benefit. Another major goal was to determine how generalizable our results are to geothermal types other than water-dominated and to other geothermal development locations. We believe that most of our research has implications for other resource types and other sites.

THE IMPERIAL COUNTY PLAN

This book focuses on research to assist in the energy planning process at the county level. In view of the model for energy planning and development just presented, it is important to ascertain the extent of adherence to the model by the study region, Imperial County. Such an evaluation at this early stage in the research means that our research conclusions will be presented in advance of the readers' understanding of them. This is deemed reasonable, however, since we are not discussing the substantive content of the conclusions but merely listing them and seeing if they are adhered to in county planning.

Adherence to the research and policy suggestions can only be evaluated after geothermal plans come on line from 1983 onwards for a period of 25 to 50 years. At this time point, we can only ask how much of our research was incorporated in the county's General Plan—an incorporation which may or may not imply actual implementation.

Imperial County's General Plan consists of the following designated series of elements, most of which have been completed, as of summer, 1980:

1. Conservation
2. Open space
3. Leisure safety
4. Recreation
5. Noise
6. Agricultural
7. Land use
8. Circulation
9. Housing
10. Geothermal

Some, such as the conservation element and noise element, are mandated by the state; others, including the recreation and geothermal elements, are optional. Most elements were written by the Planning Department. The geothermal element was published in December 1977, after adoption by the County Board of Supervisors on November 22, 1977. (County of Imperial, 1977)

The question of extent of adherence of the General Plan to the model will be answered in two parts. First, the scope and level of sophistication of planning in general and of energy planning in particular will be discussed relative to the model. Next, the research inputs from our research project to the geothermal element will be examined.

SCOPE OF PLANNING AND OF ENERGY PLANNING IN IMPERIAL COUNTY

As a small rural county, Imperial County cannot be expected to use the modern data-analytic techniques for planning proposed by the model. The county has detailed goals in the areas of agriculture, planning, community development, economics, environment, housing, health, law and justice, social services, transportation, and county administration.

The following goals are listed for comprehensive planning:

1. Formulate county's General Plan.
2. Perform zoning in consistency with General Plan.
3. Regulate the design and improvement of subdivisions.
4. Develop plans for unincorporated communities.
5. Manage growth by keeping development around cities and away from prime agricultural land.
6. Revise planning considerations for old undeveloped subdivisions.

7. Create an action plan to develop the Salton Sea as a recreational area.
8. Increase knowledge of geothermal resources. Encourage beneficial multiple use of geothermal resources. Develop geothermal resources compatible with agriculture and the environment.
9. Provide sewers. Assist financially and otherwise with other public improvements.
10. Provide for flood protection and control.

This list is revealing in hinting at geothermal as a unique area of the county plan—the only area using a modern data-oriented, integrated approach. Such a supposition is further confirmed by comparing the land-use and conservative elements with the geothermal element. The land-use element was approved by the County Supervisors and published on June 25, 1973 (County of Imperial, 1973). After reviewing existing conditions affecting land use, the report devotes eight pages to planning land uses. The 15 points of land use policy mentioned are conventional, focusing on avoidance of conflict between industrial and prime agricultural land uses. Next, eight types of land uses are defined. Finally, several paragraphs are devoted to the interactions of land uses with geothermal development, transportation, and water resources. This document reveals a mundane land-use policy lacking in modern elements.

The conservation element was passed and published on December 18, 1973 (County of Imperial, 1973). The report is organized by type of resources—water, land, minerals, soils, agricultural lands, energy resources, geothermal resources, and biological resources. The plan's goal is a compromise—"the goal . . . is not to return to a pre-industrial rural lifestyle nor to develop all our available resources, but rather to develop and implement a sensible plan for their utilization and conservation." An example of a planning step conforming to this goal is the installation of drainage tiles in the agricultural fields to prevent salt accumulation in the soil. Unlike the land-use element, the plan calls for development of a modern data base and a need to draw on all available expertise, "including Federal and State agencies as well as local governments and special districts." Universities are also designated as a vital information source—"educational institutions should be encouraged to promote research projects and studies relevant to the information needs of the County."

The individual sections of the report tend to be comprehensive, but fairly commonplace in approach without a great deal of specificity to the county. The geothermal resources section, for example, consists of a review of conventional facts about geothermal energy. Only a few of the elements of the model are called for—geotechnical and environmental planning requirements are mentioned, but excluded are planning considerations for extralocal conditions, socioeconomic influences, and political systems and outcomes. In general, this planning element is an improvement on the land-use element, but it still falls short of the model's sophisticated techniques and broad, integrated approach.

The geothermal element incorporated the multidisciplinary research results of the University of California, Cal Tech research project—the authors' socioeconomic results discussed later in this book, as well as the results of the other investigation portions of the research including geoengineering, economics, environmental, etc. It is not surprising that the geothermal element is much broader and more modern in approach than the other elements. The geothermal element's contents include the following:

(1) Appraisal of research methodology. This is a vital part of the model and is not contained in any of the other elements.

(2) Definitions of the planning process. The planning process includes the following steps: (a) data collection, (b) analysis of existing conditions, (c) problem definition, (d) forecasting, (e) establishment of goals, (f) establishment of public input, (g) determination of alternatives, (h) procedures for achievement of goals based on political and technical criteria, and (i) adoption and implementation of plan. Steps a, b, d, and h are parts of the model. Moreover, step f would surely be beneficial (see Chapter 7 for more discussion of this point). Overall, the planning process is modern and integrative. It differs greatly from the county's 1973 comprehensive planning outlined earlier.

(3) Background on geothermal energy and a description of Imperial County.

(4) Summaries of all parts of the UC Riverside—Cal Tech research projects. Material covered here covers much of that called for in the model. Model areas not covered include the cost of energy from alternative sources, meteorological data, certain other physical environmental data, etc.

(5) Proposals for future research projects. Clearly the county has seen the value of a modern approach and is calling for additional research. Projects called for include a master transmission corridor study and a study of the most feasible uses of direct heat from geothermal resource.

(6) The Master Plan for geothermal development. A condensation of the researcher's conclusions and proposals.

(7) County administrative assignments for carrying out the plan.

The geothermal element was farsighted and bold. In subsequent chapters we note which of our research results were included in the county's plan and which were excluded.

In summary, Imperial County has made a good start on use of the Research Model in the geothermal element of the County's General Plan. However, in other areas of its General Plan, it has not used modern approaches equivalent to that of the model. One weakness in the geothermal element is the restriction to a scope somewhat narrower than outlined earlier for the model. Another weakness is exclusion of some of the research and policy results.

The plan will likely be revised as geothermal development unfolds. Clearly the plan should be broadened. Possibly a new comprehensive research study could be commissioned in say, 15 years, to update the plan. Overall, however, we feel that the county's boldness in integrating the research, accepting the concept of a broad model, and using substantial research results is unusual and commendable.

We recommend that other counties anticipating geothermal development adopt our research model at the earliest possible stage in the geothermal development process. Much can be learned from examining the Imperial County experience with planning and development. However, such a county should be very wary of using specific research conclusions, without localized data.

Acknowledgment and Comment

Dr. Martin J. Pasqualetti helped formulate parts of this chapter; his help was of great benefit to us! For further explanation on points made in this chapter, see Pasqualetti, Pick, and Butler (1978).

2

Geothermal Energy in Imperial County, California

GEOTHERMAL RESOURCES

Definitions

Geothermal energy is the natural heat of the earth. Anyone who has witnessed a volcanic eruption or penetrated the interior of a deep mine will attest to the observation that temperature rises with depth in the earth. This heat of the earth everywhere flows to the surface by conduction and by flowing water. Geothermal energy becomes a *geothermal resource* when it is capable of economic utilization. Geothermal energy is of economic interest only where heat is concentrated above the normal background level of the earth's natural heat flow.

Underlying Geological Theory of Geothermal Energy

Recently developed geological theories have shed light on the origin and occurrence of geothermal resources. Although these theories are to some extent speculative, because of the impossibility of direct observation of the processes postulated in operation, there is widespread acceptance of the theory of *plate tectonics* and fair agreement on theories of *geothermal reservoir formation*.

The theory of plate tectonics has been firmly established only since the mid-1960s. It stems from the old geological theory of continental drift. Pre-twentieth-century geologists noticed the close fittings of the shapes of the eastern and western coastlines of the Atlantic Ocean. They speculated that the coastlines may have been joined together in ancient times. Beginning in 1912, Alfred Wegener and his colleagues gathered many types of evidence suggesting an ancient supercontinent, Pangaea, which split apart and drifted to form other continents.

Modern tectonic plate theory refines the old continental drift theories. According to tectonic plate theory, the surface of the earth acts like a series of separate plates moving in different directions. The rate of such movement is several centimeters per year. The plate motions involve the upper layer of the earth called the lithosphere, as shown diagrammatically in Figure 2–1. These plates are not limited to the continents and continental shelves—the 10% of the continental masses which are below sea level—but also include the earth's crust beneath the oceans (Figure 2–1).

The plates of lithosphere move in different directions as if on conveyor belts, by convection of the deeper asthenosphere layer. At their boundaries, the lithospheric plates on the two sides may diverge, as in the middle of the Atlantic Ocean; converge, as along the western edge of South America; or move parallel to each other, as along the San Andreas Fault.

As seen in Figure 2–1, it is at the plate boundaries that lava from the earth's asthenosphere usually rises, at times reaching the earth's surface to form volcanoes. When lithospheric plates move towards each other, one plate buckles beneath the other, as seen at the western edge of South America.

A simplified version of the world's plate boundaries is shown in Figure 2–2a. The divergence of tectonic plates under the ocean used to be referred to as "sea floor spreading." In addition to volcanism, plate boundaries may be characterized by young, active mountain ranges, ridges on the ocean floor, earthquake activity (seismicity), volcanoes, and reservoirs of geothermal energy in various forms. The young mountain ranges, such as the Himalayas, are hypothesized to have been formed by the buckling of lithosphere at plate edges. The mid-ocean ridges are the result of the movement apart of plates. As drifting apart takes place, fault lines may be formed perpendicular to the ridges to relieve stresses. Earthquakes are assumed to express release of stresses which have built up in the earth's layers. They are characteristic of the lithosphere rather than the asthenosphere, because the asthenosphere is not rigid enough to support brittle deformation necessary for earthquakes.

Major evidence for the support of tectonic plate theory stems from the earth's history of magnetic polarity reversals, paleomagnetism, and borings of the ocean floor from the National Science Foundation's Deep Sea Drilling Project (Wyllie, 1976). Magnetic evidence is utilized because magnetic specimens of rocks retain the magnetic patterns in-

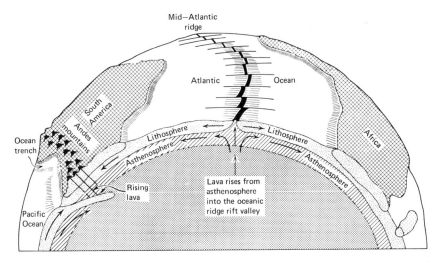

FIGURE 2-1. The geology of tectonic plates showing sea-floor spreading and continental drift. (Source: Wylie, 1976.)

duced at their time point of origin. The deep sea borings are valuable in confirming the movement of tectonic plates away from the deep ocean ridge mentioned above. This movement is deduced because the age of the sediments increases with distance from the ridges and the thickness of the sediments also increases, indicating that the sedimentation rate does not change with distance from the ridge.

Finally, most geothermal resources are found at plate boundaries, because of the higher heat flow and volcanicity which occurs there. Shallow intrusions of molten rock may heat the interstitial water to form geothermal reservoirs. A plot of the world's geothermal regions (Figure 2–2b) shows the close geographical correspondence of these regions and the plate boundaries (Figure 2–2a). For example, Iceland lies on the boundary between the North American and Eurasian plates. However, it is clearly evident from Figure 2–2b that there are some exceptions to this correspondence, such as Hawaii. For those areas with close correspondence, complex, historical geological patterns determine a somewhat different relationship between plate tectonics and geothermal deposits at each site. The geothermal resources of Imperial County are a manifestation of its position on a plate boundary, in which tectonic plates are moving parallel to each other in a northwest to southeast direction (Figures 2–2a, 2–2b, and 2-2c, (Elders *et al.*, 1972). One of the effects of this plate movement is geologic faults, referred to as transform faults, which are the southern part

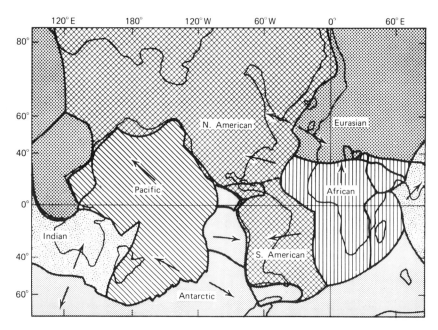

FIGURE 2-2a. The world's tectonic plates and plate movement. Much of the world's geothermal resource occurs around the edge of the Pacific Plate, sometimes referred to as the Rim of Fire. (Source: Wyllie, 1976.) (b)

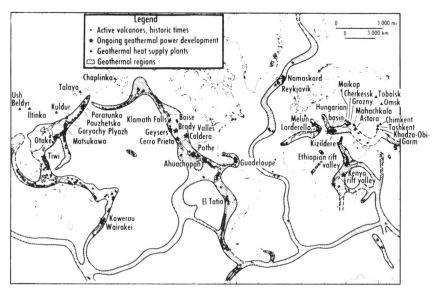

FIGURE 2-2b. The world's geothermal regions. Most are located at tectonic plate boundaries. (c)

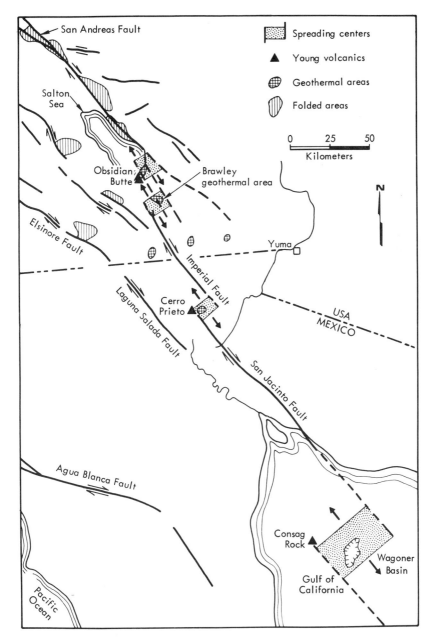

FIGURE 2-2c. Spreading centers and geothermal areas in Imperial County. The centers are associated with tectonic plate boundaries. (Source: Elders, 1972.)

of the San Andreas fault system. These include the Imperial Fault, the San Jacinto Fault, and the Elsinore Fault, as well as the San Andreas Fault itself. Perpendicular to the transform faults are certain regions in which crusted spreading is occurring, known as spreading centers. These are analogous to the sea floor spreading, previously referred to. They are represented in Figure 2–2c by the dotted areas.

In the course of geologic history, the Imperial County spreading centers are hypothesized to have been characterized by magmatic intrusion, metamorphism, sedimentation, and widening of centers parallel to the faults. Eventually, heating of water led to the geothermal convection systems which presently characterize the spreading centers (Elders et al., 1972).

In summary, regions located on or near tectonic plate boundaries are characterized by potential for geothermal energy. Geologic circumstances have led to evolutions of actual geothermal fields in certain of these regions, such as in Imperial County.

Types of Geothermal Resources

Geothermal resources may be classified according to mode of formation and temperature. The classification of the U.S. Geologic Survey, shown in Table 2–1, distinguishes types by the manner in which heat moves upward. For hydrothermal resources, heat is carried upward from deep, hot rock layers by convection currents consisting of

TABLE 2–1. TYPES OF GEOTHERMAL RESOURCES BY GEOLOGIC MODE OF FORMATION[a]

Resource type	Temperature
1. Hydrothermal	
(a) Vapor dominated	About 240°C (464°F)
(b) Water dominated	
(1) High temperature	150–300+°C (302–572°F)
(2) Medium temperature	90–150°C (194–302°F)
(3) Low temperature	<90°C (194°F)
2. Hot igneous rock	
(a) Part still molten	>650°C (1202°F)
(b) Not molten	90–650°C (194–1202°F)
3. Conduction dominated	
(a) Geopressured	150–200°C (302–392°F)
(b) Radiogenic	30–150°C (86–302°F)

[a] Source: USGS Circular 726, 1975, modified by Wright (1978).

FIGURE 2-3. Steam release at The Geysers, Sonoma County, California.

either vapor (vapor-dominated resource type) or hot water (water-dominated type). In the case of vapor-dominated resources the vapor, such as that shown in Figure 2-3, which is almost entirely steam, condenses near the surface, except for occasional surface release in geysers. Condensed vapor may seep back down to a lower zone of vapor to be reheated into steam. For both hydrothermal resource types, natural fractures may be necessary to allow *permeability*—that is, capacity for transmitting fluids through rocks under a given pressure gradient. Related to permeability is *porosity*, or the ratio, in rock, of pore volume (nonrock space) to total volume. As the hot fluids convect upward, there are losses to the geothermal reservoir. Continuity of mass requires that the reservoir be recharged by colder waters. Geophysical explanations of mechanisms for recharge are currently speculative.

For the water-dominated resource, there is a large temperature range. Medium- and low-temperature resources may occur at the

FIGURE 2-4. A geothermal flash steam turbine at Cerro Prieto, Mexico.

edges of high-temperature geothermal fields or in separate areas, often shallow in depth, for which heat flow is too low or convection is too high to cause high temperatures. Geothermal electrical generation is presently only possible for the high-temperature category through the use of a steam turbine as shown in Figure 2-4. However, there are many direct use applications of the heated water for all three temperature ranges. Several of Imperial County's direct use applications are covered in detail in Chapter 8.

The second type of geothermal resource, hot igneous rock, is based on direct heat flow in molten or solidified igneous rock. The molten subtype involves magma, that is, molten rock, at temperatures above 650°C (1202°F). The hot nonmolten subtype, commonly referred to as hot dry rock, is currently being studied in field experiments at Los Alamos. In the experimental procedure, a well is drilled into hot, tight igneous rocks to a depth of about 3 km (10,000 ft). Water is pumped down the well causing artificial fracturing of the deep rock layers. A second well is then drilled to a depth of about 2 km (6,000 ft). Water pumped through the deep well is percolated upward through the rock, heated, and retrieved as a hot water geothermal brine by the shallower well. Hot dry rock generation of electricity has too many technological problems to be commercially feasible yet.

The third category of geothermal resource is conduction dominated. In this case, heat is carried upwards by conduction through rock layers. For the geopressured subtype, hot water, which is at depths up to 9 km (30,000 ft) and pressure greater than hydrostatic pressure, is the conduction medium. Often a substantial amount of methane is contained in a geopressured deposit. The energy potential from a geopressured resource is threefold: the conventional (hydrothermal) geothermal energy in the hot water, the mechanical energy stemming from the high pressure, and the fossil energy contained in the methane gas. The Department of Energy is conducting field geopressured research in Louisiana; presently this resource type is not commercially feasible.

A second subtype in the conduction-dominated category, the radiogenic subtype, receives heat from decaying radioactive elements located in granitic rocks and is located mainly in the Eastern United States. Research has been conducted at the Virginia Polytechnic Institute on this type, but it is not useful for commercial exploitation as yet.

Geothermal Exploration and Drilling

If a region is suspected of possessing a commercially exploitable geothermal deposit, the initial steps in resource utilization are exploration and drilling. Drilling towers like the one in Figure 2-5 are similar to oil drilling rigs. The discussion in this section pertains to water-dominated hydrothermal resources, although some methods apply also to vapor-dominated hydrothermal and hot dry rock.

Although location on a tectonic plate edge and/or surface manifestations, such as geysers, hot springs, fumaroles, etc., increase the probability of a geothermal reservoir, it is usual to employ a series of geophysical, geological, and geochemical techniques. Geophysical techniques employed include gravity, heat flow, seismics, and electrical resistivity. After these techniques have been applied to a potential resource area, a series of maps may be drawn indicating high temperature or other anomalous values of geophysical parameters associated with high heat flow. Because these different approaches measure different geophysical parameters, the contours of, say, the anomalous magnetic measurements do not coincide with the gravity or electrical anomalies. Usually, more than one technique is necessary to achieve confidence on reservoir shape and size. The rationale of these techniques is discussed below, while specific application in Imperial County is discussed later in the chapter.

Gravity is measured with gravimeters, sensitive to a gravity

FIGURE 2-5. Drilling tower at Union Oil's North Brawley geothermal field.

charge of one part in ten billion. In some instances, higher gravity can indicate the presence of a geothermal reservoir, because hot water tends to deposit minerals in the rocks, increasing density slightly and, hence, increasing the gravity reading at the surface. Heat flow anomalies are determined by temperature gradient measurements in test wells. Much more important than gross temperature is the rate of change per unit of depth—the temperature *gradient*. Sometimes the shallow temperature gradient can be extrapolated to estimate temperatures at depths of 1.5–3 km (5,000–10,000 feet). Earlier in the chapter, tectonic plate edges were characterized by the presence of both seismicity and geothermal deposits. Earthquake activity often aids in indicating geothermal energy, as has been the case in Imperial County. Usually, for this method, especially sensitive recording seismometers are planted around a suspected geothermal area for several months.

Magnetic anomalies are caused by changes in the magnetic susceptibility in the subsurface. A source of error is that such changes are confused with nongeothermal magnetic sources. The last technique, electrical resistivity, aims to measure variations in resistivity related to the presence of brine or of elevated temperatures. It is based on the electrical equation $E = IR$, where E is electrical current, I is voltage, and R is resistance. The electrical currents are those flowing through the surface layer of the earth, and may be related to the depth and size of geothermal deposits. Voltage and current are measured to indicate resistivity (a component of resistance).

Geological approaches to exploration operate at two levels. First, broad-scale regional studies to locate areas of good potential are necessary. These are followed by detailed site-specific studies of the surface and subsurface geology. Information from drilling is used to understand the subsurface stratigraphy and structure and to formulate a conceptual model of the reservoir. Studies of rock samples recovered during drilling are also used to investigate reactions between the rocks and the geothermal fluids, which have dramatic effects on their properties.

Geochemical techniques have also proved to be very cost-effective exploration tools. Sampling of fluids and gases from hot springs and geysers gives useful information on subsurface temperature and chemistry. For example, the concentrations of silica, as well as of calcium, and sodium to potassium concentrations are found to be useful geothermometers.

After a geothermal field has been mapped by several or all of the above methods, shallow test wells are drilled to gather production data, such as temperature gradients, pressures, chemical composition of brines, and borehole samples. Well tests and, to a lesser extent, borehole samples are used to estimate permeability—which is an important factor in the ability of a geothermal reservoir to produce. Drilling operations are performed by the same type of portable drill rigs used in the oil and gas industry. A rig is generally operated by a crew of over a dozen workers, working continuous shifts. Geothermal pipe varies in size, averaging about 9 in. in diameter. It is placed into the well hole one segment at a time.

Once fluid production commences in a geothermal field, geothermal reservoir engineering is often performed to extract more optimally the resource. Analogous engineering techniques have been used in the oil and gas industries for many decades, so that a present-day fossil fuel deposit is often scientifically "mined." Reservoir engineering techniques for geothermal energy, many adapted from the oil and gas industries, have been developed largely since the mid-1920s. The goal of reservoir engineering is to model accurately and predict production results for the entire period of prospective production. Such data are often used by financial institutions in decisions on geothermal financing. The techniques in use are (1) volumetric, (2) performance, (3) numerical modeling, and (4) new methods. For (1), the volume of brine flow from each well is measured. Flow data are important for power plant design and reinjection considerations. The second method uses decline curve analysis and material–energy balances—old methods drawn from the oil and gas industries.

Decline curve analysis is based on measurements of production rates, after production commences in a field. A plot is done of the logarithm of actual production rate versus time. Based on an appropriate theoretical model, a curve is drawn which fits the points on the plot. The curve is extrapolated for the duration of expected production. By this technique, the remaining field production and producing rates may be estimated. The procedure is continually updated, as additional performance data become available.

For material–energy balance, the principles of conservation of mass and energy are applied to production data, to yield estimates fo the total mass of the reservoir as well as estimates of future performance. The third method, numerical modeling, uses computer simulation. Computer programs are written encompassing fluids in one or more phases, rock layers, energy content, and so forth. Newer and experimental techniques include use of gravity measurements and radioactive tracers to estimate remaining production.

Besides accurate understanding of a producing reservoir, another goal of reservoir assessment is enhanced recovery. This means that the reservoir contents are scientifically extracted with injection of surface water at carefully chosen locations—an approach in widespread use in the oil and gas industry.

Geothermal Power Plants

If the political, environmental, and economic factors permit and if financing can be obtained based on exploratory and reservoir assessment data, a geothermal field development will progress into the stage of electrical production from power plants. The power plant shown in Figure 2–6 is the geothermal plant in Cerro Prieto, Mexico, not too far below the border from the Imperial Valley KGRAs. The power plant design is critical in the success of a project. It needs to be tailored to resource type, brine characteristics, reinjection requirements, environmental restrictions, surface water constraints, and so on.

FIGURE 2-6. Steam venting from production well, Cerro Prieto, Mexico.

A diagram of the most common power plant design for water-dominated resources, a single-flash system, is shown in Figure 2–7. This design has been used in the majority of existing geothermal plants worldwide, including the Italian, New Zealand, and Mexican plants. The technology is well developed and thoroughly tested and debugged. In this system, high-temperature, high-pressure brine is run through a separator, which flashes about 20% of the flow into steam. The remaining water portion is mostly sent to an evaporation pond, although some additional steam is released along the way. The destination of this water is not fixed to an evaporation pond. At Waikakei, for instance, the water is released to a stream. In Imperial County plants, it will be reinjected back into the geothermal reservoir.

The separated steam enters one or more specially designed turbines attached to electrical generators. The cooled and depressurized turbine residual steam flow is condensed to water and sent to cooling towers. A cooling tower portion is not a necessity in plant design. However, through thermodynamic principles it serves to greatly increase the turbine's efficiency. Hence, it is included for economic reasons. Some cooling tower waste is evaporated into the atmosphere. The rest is sent to an evaporation pond or reinjected.

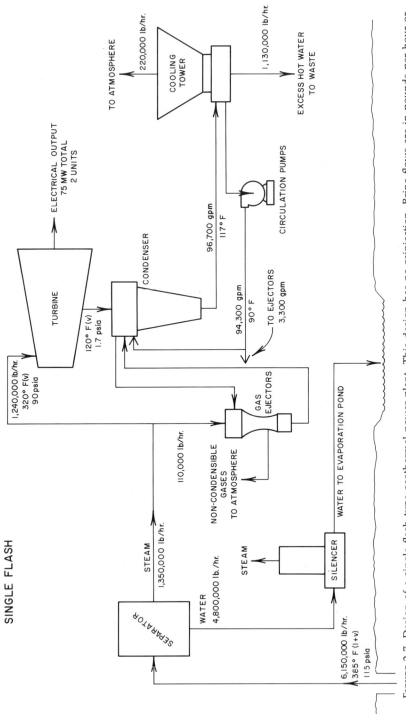

FIGURE 2-7. Design of a single-flash-type geothermal power plant. This design has no reinjection. Brine flows are in pounds per hour or gallons per minute (gpm). Pressure values are in pounds per square inch (psia). See text for explanation. (Source: VTN, 1978.)

FIGURE 2-8. Transmission lines at Cerro Prieto geothermal power plant, Mexico.

There are several important environmental considerations in plant design. First, there are many noncondensable gases, such as hydrogen sulfide, contained in the steam flow. These are released through gas ejectors, which may or may not have special air pollution control equipment (scrubbers) added. The most troublesome noncondensable is foul-smelling, hydrogen sulfide (H_2S). Another environmental problem is noise. Noise not only issues from the power plant, but also from the geothermal well siting area surrounding the power plant. In the latter, Leitner (1978) has identified the following noise sources: site preparation/road construction, geothermal well clear-out and testing, geothermal steam venting from wells, construction of steam pipelines and transmission lines (see Figure 2–8), and vehicular traffic. At The Geysers geothermal field, the worst noise levels are from well drilling and venting of steam from wells. For the latter, Leitner cites noise levels as high as 100 to 125 dBa at 20 to 100 ft. In the power plant itself, there are three major noise sources: the turbine/generator building, the steam vent gas ejector, and the cooling tower. Noise control of steam venting may be achieved by silencers. EPA has suggested maximum average day/night residential noise levels (Ldn) of 55 dB and a maximum level in open space of 70 dB. These criteria were adopted by the Lake County General Plan, California, and apply to part of The Geysers geothermal field.

Another major environmental consideration is the requirement in many geothermal plants for a surface source of water for the cooling tower system. If greater than 80% reinjection of geothermal fluids is not required, ordinarily water from the condenser is adequate to meet cooling tower design requirements. If, on the other hand, 100% reinjection of fluids is required, extra water is required from an outside source in order to maintain an adequate water level in the cooling tower's water loop. Figure 2–9 shows the same single-flash system with a surface supply of make-up water added to achieve 100% reinjection. An alternative to adding make-up water to the cooling water loop and reinjecting condensate, is to reinject the make-up water, leaving the cooling water loop alone, as in Figure 2–10. In either case, supplemental water is required. Table 2–2 shows typical make-up water values (Goldsmith, 1976a).

The environment is affected by this water demand, because make-up water must be diverted from some point in the existing natural and artificial regional water cycle, causing water loss and/or diversion problems, such as pumping and distribution canals. Figure 2–11 shows an irrigation canal near the Brawley KGRA. Additional distribution canals, as a result of the geothermal power plants, will be necessary. The environment may also be affected by the gas content of make-up water, as gasses may be released into the atmosphere by the cooling towers. On the other hand, the cooling tower has specifications for maximum levels of minerals allowed in make-up water entering the cooling water loop. The make-up water requirements and solutions for Imperial County are discussed in detail in Chapter 8.

The single-flash system is not the only possible power plant design for water-dominated resources. There are also double-flash and binary plants. A double-flash system is basically similar to single flash. In a double-flash design, however, the water exiting from the separator is flashed a second time to obtain further steam for flow to the turbine. In a binary plant design, shown in Figure 2–10, heat from

TABLE 2-2. TYPICAL MAKE-UP WATER VALUES

Reservoir temperature (C°)	150	200	250	300
Make-up water $\left[\frac{\text{acre ft.}}{\text{yr MW}_e}\right]$	97	61	50	41

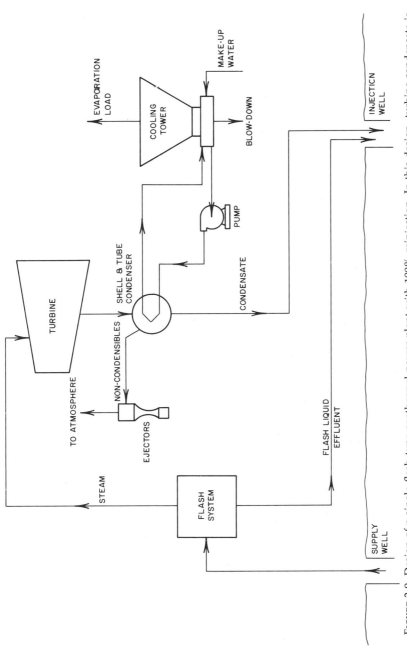

FIGURE 2-9. Design of a single-flash-type geothermal power plant with 100% reinjection. In this design, turbine condensate is reinjected into the geothermal reservoir and make-up water is added to the cooling water cycle. (Source: VTN, 1978.)

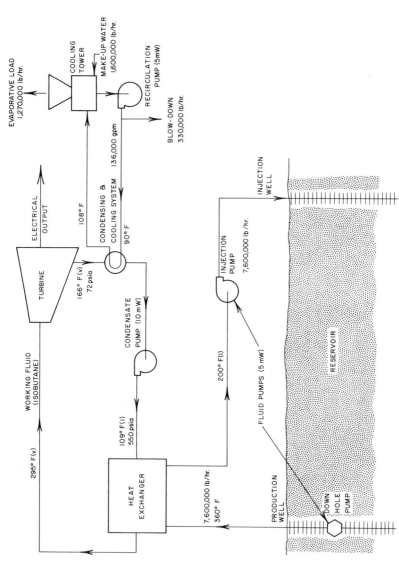

FIGURE 2-10. Design of a binary-type geothermal power plant. This design has reinjection. Brine flows are in pounds per hour or gallons per minute (gpm). Pressure values are in pounds per square inch (psia). See text for explanation. (Source: VTN, 1978.)

FIGURE 2-11. Irrigation canal near Brawley, Imperial County, California, carrying Colorado River water.

the geothermal fluids is transferred to a separate cycle of a special working fluid, usually butane. Although binary plants avoid many corrosion and scaling problems of flash systems, they are generally more expensive and require greater amounts of cooling water. The 11-MW_e (megawatts of electrical capacity) East Mesa binary plant constructed by Magma Power Company is discussed in greater detail in Chapter 8.

Thus, geothermal energy is a form of energy directly utilizing heat from the earth. The heat may be brought to the surface as steam or hot water, as shown in Figure 2-12. Geothermal development consists of the exploration and drilling of wells (somewhat equivalent to oil drilling) down to the depths of the hot water or steam, transporting the hot water or steam to the surface, and utilizing the resource to turn turbines in a power plant and generate electricity. Thus, the engineering procedure to extract this resource is to drill down to the resource, bring it to the surface, and use it to turn turbines and generate electricity. Alternatively, the hot water or steam may be used directly for a variety of applications, including house heating, air-conditioning, salt extraction, industrial process heating, or even bathing at geothermal spas—nonelectrical uses (direct uses). These are discussed in Chapter 8.

The energy potential of geothermal resources in the U.S. is hundreds of times greater than its current use. Geothermal reserves in the western U.S., as shown in Figure 2–13, have been estimated to be adequate to supply all new electric energy capacity in those states for the next two decades. This energy could be generated from natural

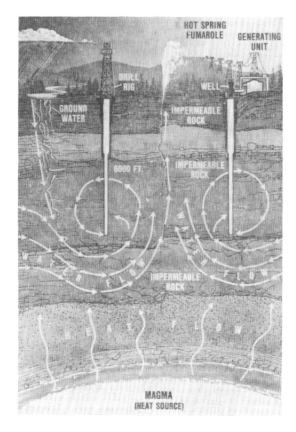

FIGURE 2-12. Magma power-volcanic activity from the earth's interior heats ground water which has seeped into impermeable rock. After drilling wells as deep as 10,000 ft, geothermal plants can harness the power to generate electricity or provide heat. (Courtesy of *Los Angeles Times.*)

steam and high-temperature geothermal fluids at prices competitive with electricity from fossil and nuclear power plants (Sacarto, 1976). In addition to generating power, geothermal resources could also be used in heating and cooling buildings, as well as in agriculture and industrial processing.

With vigorous development, geothermal resources could supply the U.S. with somewhere in the range of 10% of its electrical energy by the year 2000 with even a greater portion being provided subsequently. In comparison to other major energy sources, there has been relatively little research and development done on locating fields, locating high-temperature reserves, or in the development of technology

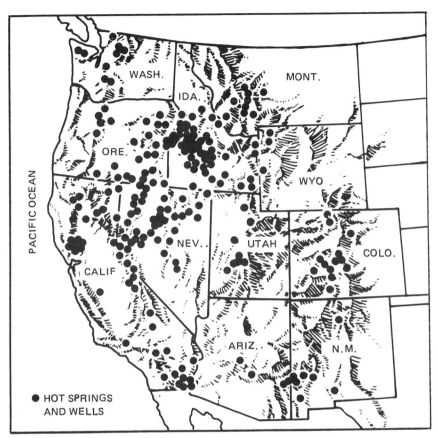

FIGURE 2-13. Surface hints, such as hot springs and wells, show potential geothermal power sources abound throughout the west. (Courtesy of *Los Angeles Times*.)

necessary to exploit hot dry rock, intermediate-temperature systems, and near-surface magma chambers.

The development of geothermal resources is relatively environmentally safe, especially when compared to coal combustion and nuclear fission, yet with a comparable, if not definitively lower, cost. Geothermal resources can be produced with waste water disposed of by injection into wells with limited disturbance to land, ground, and surface water resources. Currently the Cerro Prieto geothermal power plant in Mexico discharges its waste fluids into nearby ponds (see Figures 2–14 and 2–15). This, of course, stands in contrast to the surface mining of coal and uranium which disrupts large land areas and wildlife habitats, and may produce extensive water and air pollution (Sacarto, 1976).

FIGURE 2-14. Well-site effluent at Cerro Prieto, Mexico.

There are some serious obstacles to the development of geothermal resources in the U.S. One, of course, is that investment must be greatly expanded beyond its current level. One estimate suggests that $75 billion will be needed to discover and produce 100,000 MW_e of electric power from geothermal resources; this means that approximately 65,000 wells will have to be drilled. At the present rate of development, this goal will not be reached for several centuries. Clearly, electrical utilities are very aware of this relatively unused and unfamiliar energy resource. A study done at the Stanford Law School stated that "although estimates of geothermal potential are very speculative at this time, it is apparent that potential is greater than the extent to which the California utilities care to explore it" (cited in Ternes, 1978:74). In the U.S., at the present time, there is only one set of large-scale geothermal plants located at The Geysers area in northern California. This is the largest set of geothermal plants in the world.

Regulations established for the utilization of fossil fuels and in nuclear fission generation of energy are not suitable for the development of geothermal resources; this is clearly a substantial problem. Federal energy research and development financing has also failed to

FIGURE 2-15. Brine discharge into evaporation pond, Department of Energy test site, East Mesa.

encourage large-scale geothermal development and generally has treated it as an exotic phenomenon to be studied rather than employed (Sacarto, 1976:2). Generally, almost all federal, state, and local policies accentuate the uncertainty and risk inherent in geothermal exploration, research, and development rather than its great energy potential with relatively little environmental degradation.

The most prominent and most clearly identified geothermal resources in the U.S. currently occur in the 15 Gulf and western states. Generally, in each of these states, an authority and guidelines have been established to administer geothermal leasing and other aspects of development. According to Sacarto (1976), these policies address the following:

(1) *Resource definition:* a description of the resource subject to geothermal policies.
(2) *Leasing provisions:* specifying the manner of leasing and lease obligations.
(3) *Development regulations:* the control of well drilling and production for purposes of safety and resource conservation.
(4) *Water appropriation:* to fix the adaptability of state laws to geothermal development.
(5) *Environmental standards:* regulating the permissible effects of geothermal development on the environment.

There are other major issues that could have serious consequences for geothermal development; in general, they have not received the attention of policy makers. These include the following:

(1) *Taxation:* property, severance, income, and other special taxes influence geothermal development.
(2) *Securities regulations:* state securities commissions and the Federal Securities and Exchange Commission can expand or contract the supply of capital available to geothermal developers.
(3) *Utility regulations:* utility investments and power transmission are two important policy areas within utility commission jurisdiction.

Each of these areas generally has established Federal, state, and local policies, many of which currently are not responsive to the needs of geothermal development. Generally, established policy now presents obstacles to geothermal development rather than encouragement.

The following conditions are needed for the geothermal industry to pursue large-scale development: (1) consumer (utility) confidence in the resource; (2) equitable tax treatment; (3) prompt exploration of extensive land areas; (4) long and secured tenure for productive properties; (5) prompt facility siting and development; (6) competitive access to various consumers (Sacarto, 1976:3).

USE OF GEOTHERMAL RESOURCES

Geothermal resources have been known and used for centuries. Hot mineral springs have been visited for their therapeutic and recreational value. However, during the past several decades, developing technology has made geothermal energy available for energy purposes. In 1976, geothermal power plants throughout the world generated approximately 1400 MW_e of electrical power. In 1979, this worldwide geothermal power plant capacity was about 2500 MW_e, including plants coming on-line shortly. Present (1976) and estimated future installed capacities are presented in Table 2–3 by country of occurrence. In addition, in some countries, geothermal resources are used for direct heat, including the Soviet Union, Hungary, Iceland, and New Zealand, as well as in a minor way, the U.S. These countries, in combination, use approximately 6000 MW_t (i.e., thermal megawatts) in agriculture, in industrial processing, and for heating and cooling of buildings. In the U.S., The Geysers, a dry steam field north of San Francisco, provides slightly over 500 MW_e of electric power, or enough to supply a city of one-half million population. Currently, geothermal energy is used directly for heating in scattered locations in the U.S. such as Klamath Falls, Oregon, and Boise, Idaho.

TABLE 2-3. WORLDWIDE GEOTHERMAL GENERATING CAPACITY AND FUTURE PROJECTIONS[a]

Country	1976 Installed capacity (MW)	1985 Estimated capacity (MW)	2000 Estimated capacity (MW)
United States	522	6,000	20,000
Italy	421	800	—
New Zealand	202	400	1,400
Mexico	78.5	400–1,400	1,500–20,000
Japan	70	2,000	50,000
El Salvador	60	180	—
Soviet Union	5.7	—	—
Iceland	2.5	150	500
Turkey	0.5	400	1,000
Phillipines	—	300	—
Nicaragua	—	150–200	—
Costa Rica	—	100	—
Guatemala	—	100	—
Honduras	—	100	—
Indonesia	—	30–100	500–6,000
Panama	—	60	—
Taiwan	—	50	200
Kenya	—	30	60–90
Portugal	—	30	100
Spain	—	25	200
Argentina	—	20	—
Canada	—	10	—
Total	1,362.2	11,335–12,455	75,460–100,000

[a] Source: Meidav, Sanyal, and Facca (1977).

In terms of its real and already developed potential, the current use of geothermal energy is relatively small. The total source of geothermal energy, or energy flowing from the earth's core to the earth's surface, is an amount of energy generated annually at a rate equivalent to at least ten times the 1974 human world energy consumption. The ultimate source of geothermal energy is nuclear processes deep in the heart of the earth. "Radioactive thorium, potassium, and uranium, distributed more or less evenly in the earth's crust yield heat, as they decay" (Sacarto, 1966:7). These elements are long-lived and have half-lives of several billions of years and apparently this heat will continue without fail for a millennium.

Despite the substantial store of energy beneath the earth, its practical value depends on an ability to utilize it effectively. Geothermal resources, like other resources, are most valuable when they are near the earth's surface. The resource must be hot enough for it to be a practical and economical resource. Thus, many sites are not useful for commercial utilization. Nevertheless, many valuable sites have been found in the U.S. and elsewhere throughout the world.

Generally, the basic structure of geothermal resources, where they have economic potential, reflect the basic geological structure of the planet. One basic geothermal region follows the Pacific shoreline, all the way from the Aleutians, through Alaska, down through Canada and the western U.S., on down through Central America and the Pacific coast of South America. This particular geothermal region is linked to others that flow up through the middle of the Atlantic Ocean between the continents and through the area on the other side to the perimeter around the Pacific Ocean. Thus, the whole western U.S. is a basic geological area in which geothermal resources are economically viable.

Geothermal Resources in Imperial County

> A realistic assessment of the geothermal resources in Imperial Valley is a necessary prerequisite to the development of potential geothermal areas in the estimation of its impact on the economy, environment, and society. Several uncoordinated efforts toward the assessment of the geothermal energy in the Imperial Valley have been made by various government agencies, universities, and private industry. Because of the lack of coordination and cooperation among the various contributions from different sectors, some of the effort has been duplicated and in some cases, there appear to be conflicting interpretations. For example, assessment of geothermal resources made by different workers may differ by several orders of magnitude, reflecting a lack of exchange of ideas and data (Biehler and Lee, 1977).

Identifying potential geothermal reservoirs is not an exact science. A number of geological and geophysical exploration methods have been applied in Imperial County with some success. The best way of determining the existence of geothermal reservoirs is to drill deep test holes. Other techniques also have been used to determine areas for deep drill tests. Only one geothermal reservoir in Imperial County has surface geologic features—the south end of the Salton Sea, where five small volcanic plugs are visible. Other potential reservoirs have been located by the subsurface geophysical methods outlined below:

(1) Gravity studies indicating that local potential geothermal reservoirs are associated with excess mass in subsurface rocks. Gravity

techniques can be used as a method for outlining areas for further exploration. Gravity exploration is relatively inexpensive and has been used extensively throughout Imperial County.

(2) Temperature profiles in shallow test wells measure geothermal gradients from which subsurface heat flow can be estimated. These data are not always clearcut indicators of a geothermal reservoir. There is a very close relationship between test wells with high-temperature gradients and those with markedly positive gravity anomalies.

(3) Seismic studies also have been used to locate geothermal reservoirs. There appears to be a relationship between seismic noise and geothermal reservoir locations.

(4) Ground and aeromagnetic methods covering the entire valley indicate that some geothermal anomalies are associated with magnetic highs (or positive magnetic anomalies), e.g., the Salton Sea geothermal area (Griscom and Muffler, 1971). Generally, data suggest that fields are much larger than surface geology indicates; thus, the gravity map of the Salton Sea suggests it is rather extensive.

(5) Electrical measurements are also good indicators of areas where temperature gradients are high; however, it is difficult to separate clearly the effects of temperature from that of salinity. Thus, a relatively cold area with high salinity could show up as an anomaly even though it would not necessarily indicate a geothermal reservoir.

(6) Aerial remote sensing shows major fault structures. However, within agriculture areas, ground has been disturbed so seriously that very little can be seen from high-altitude photographs regarding surface geology.

On the basis of all these types of measurements, several areas have been designated as potential geothermal reservoirs in Imperial County. Of these, only one, the East Brawley area, has not been designated by the U.S. government as a KGRA (Known Geothermal Resource Area). Figure 1–2 shows the boundaries of Imperial County KGRAs. Even as late as 1976, only four of these KGRAs were thought to have economic potential—Heber, East Mesa, North of Brawley, and South of the Salton Sea. San Diego Gas and Electric/Department of Energy pilot geothermal power plant at Niland is shown in Figure 2–16. The areas at Glamis (along the Coachella Canal) and The Dunes currently are thought to be too small and insignificant to be economically productive. Magma's pilot plant at the East Mesa is shown in Figure 2–17. The KGRAs shown on Figure 1-1 include about 1,000 km^2 out of a total 10,000 km^2 of Imperial County. However, this is not the real measure of geothermal potential in Imperial County. Only half of the KGRA area set aside by the U.S. government, about 500 km^2, ap-

FIGURE 2-16. Tanks at San Diego Gas and Electric/Dept. of Energy pilot plant, Niland.

pears to be economically productive. This is approximately 5% of the land area in Imperial County.

There may be other potential geothermal reservoirs in Imperial County which have not yet been delineated. In order to clearly determine that no other geothermal areas exist, an extensive program of measurements would have to be accomplished.

Origin of the Geothermal Resource

The origin of the geothermal resources in Imperial County is intimately tied up with its geological history and linked with the San Andreas Fault and spreading centers associated with the East Pacific Rise under the Pacific Ocean. The North American Plate is moving in a westerly direction away from the mid-Atlantic Ridge. About 23 million years ago, as the plate moved westward, it came in contact with

FIGURE 2-17. Tanks at Magma Power's pilot plant, East Mesa.

the spreading center of the East Pacific Rise. The Pacific Plate bent downwards and was subsumed underneath the North American Plate at the Farallon Trench under the Pacific Ocean. The bending of the plate later resulted in the San Andreas Fault (W. Elders, personal communication, 1979). Subsequently, the Trough of Imperial Valley has expanded to almost twice its previous length. There is extensive block faulting along the flanks of the valley. Probably, the major source of heat in Imperial County is in storage within brines that are heated by magmatic implacement in the crusts and portions of the lower basement (Biehler and Lee, 1977). There is some conflict over whether or not the entire valley trough is a geothermal reservoir. One argument suggests that it is not but that there are other additional areas besides the known KGRAs undergoing recent magma emplacement within the basement of the valley. Others believe that the entire valley is a vast geothermal reservoir. The source of heat is generally associated with a transform center and the heat is constantly being supplied in the lower portion of the crust and possibly within the basement.

There is some question about the emplacement of heat within the basement. If one knows the rate of emplacement in calculating withdrawal, then it can be determined whether or not the geothermal resources of the area have infinite or a definitive life span. Generally, current estimates regarding rate of withdrawal are a factor of about 10 to 100 times greater than the rate of implacement, suggesting that the geothermal resource potential in the area has a finite life span (Biehler and Lee, 1977). Rate of withdrawal, of course, is important because if the heat is more rapidly taken out than it is replaced, the system will eventually cool down and the resource will be depleted. Nevertheless, by waiting long enough (although maybe too long for realistic consideration), it will regenerate into another active geothermal system.

While current research suggests that KGRAs are highly localized, they are not always located only within the boundaries as specified. For example, there are probably many slivers and pools actively undergoing spreading at this time. This is partially related to fracture porosity and permeability. When porosity and permeability result in a highly localized geothermal reservoir, the result is one with high economic potential. Generally, on the basis of estimates that have been made with the rate of heat withdrawal to produce somewhere around 2000 MW_e in the area over a period of one year, the life expectancy of the Salton Sea geothermal fields can be estimated to be somewhere between 20 and 100 years (Biehler and Lee, 1977). More detailed calculations are not possible because of a lack of detailed technical subsurface temperature depth information and information regarding the opening and spreading of the Valley Trough.

Estimation of Geothermal Temperature and Reservoir Capacity

Estimates of the total heat and storage in Imperial County geothermal reservoirs have been discussed by a number of people. Estimates vary by four to five orders of magnitude. Generally, the earlier the estimate was made, the more inconclusive were the data and estimates. Biehler and Lee (1977) also argue that some estimates are off because of mathematical calculation errors. More recently estimates made by Towse (1976) and one by Renner *et al.*, (1975) appear to be similar although they were based on different approaches to measuring geothermal resources. The estimate of Renner *et al.* is based on simple geometry of each reservoir calculated as a product of assumed volume, volumetric specific heat of 0.6 cal/cm^3 °C, and temperature in degrees above mean annual surface temperature assumed to be 15°C; volumetric specific heat is an estimate essentially of the number of calories of heat in storage. Towse used different assumptions. He primarily used temperature gradients from shallow surface drill holes which were extrapolated to greater depths. Following these different procedures, the estimates that Towse and Renner *et al.* made fit quite closely. Table 2–4 reports other estimates of the resource.

Another technique was used by Biehler and Lee (1977), although the resource estimates are of the same order of magnitude. These estimates were primarily carried out by using a map of the gravity anomalies in the Imperial Valley and by using a similar map of temperature gradients. Using residual gravity anomalies to calculate the total excess mass for each KGRA, Biehler and Lee calculated that the geothermal resource is much the same as the Towse and Renner *et al.* studies. They did this primarily by calculating the volume of producible fluid in the reservoirs. Their simplifying assumption used an estimate of 20% for the average porosity of settlements as computed from various well logs and an estimated yield of 80% of the total fluid in the reservoir which could be utilized.

While there are a number of KGRAs currently located in Imperial County, it appears that well over half the total heat in storage is located in the Salton Sea anomaly. The Dunes and Glamis fields are apparently not large enough for commercial exploitation while the Heber, East Mesa, and Brawley ones are, but in total have less heat reserve than the Salton Sea KGRA.

Resource evaluations rely on knowledge about the volume, temperature, porosity, permeability, fluid composition, and heat capacity of the reservoir. In Imperial County, these data are inadequate; thus certain assumptions are required to estimate the extent of the resource. As new data are gathered, resource evaluation will need to be updated. As we have already pointed out, estimates by Towse (1976) and Renner *et al.* (1975) differ by a factor of 8 but seem to agree on the relative heat storage in the KGRAs. Regardless of the techniques and methods used in estimating the reservoir, the Salton Sea field contains

TABLE 2–4. ESTIMATES OF MW_e, 30 YEARS FOR IMPERIAL COUNTY KGRAs

Brawley	East Mesa	Heber	Salton Sea	Westmoreland
333[a]	487[b]	973[b]	2759[a]	1710[c]
1000[d]	477[a]	973[a]	3400[c]	
1700–2000[e]	360[c]	650[c]		
640[c]				

[a] USGS Circular 726.
[b] JPL (1976).
[c] USGS Circular 790 (1979).
[d] ERDA–DGE Scenario (March, 1977).
[e] MITRE (1977).

a little over half of the estimated geothermal resources in Imperial County.

One of the real problems of the geothermal resources in Imperial County is fluid salinity. This is particularly crucial in Imperial County since salinity increases northwestward toward the Salton Sea where most of the KGRA resource lies. On the other hand, in deeper geothermal resources, salinity is less, although data are relatively scarce. The salinity of geothermal fluid cannot be determined from the salinity of ground water (Biehler and Lee, 1977). Thus, one of the main problems of geothermal fields in Imperial County is that the Salton Sea field, the largest by far and with the highest recoverable heat content, has the poorest quality geothermal fluids. Thus, the other smaller geothermal fields appear to be those that will be developed first, but from the long-term view, the Salton Sea field has the largest potential once the salinity problem has been overcome.

So far, no hot springs have been found in the KGRAs in Imperial County with the minor exception of some mudpot activity in the Salton Sea area. Apparently, cap rocks have sealed the geothermal reservoir. This may be because of self-sealing by high-temperature fluids and water–rock interactions. This could change because the Salton Sea trough area is still tectonically active, characterized by crustal rifting and shearing. There is seismic activity in the variety of major and minor faults. Ground noise and seismicity suggest that fracturing is occurring frequently in the geothermal field (Biehler and Lee, 1977). In fact, the possibility of withdrawing hot fluids and the reinjection of cold water may change the state of the stress system, creating some possible seismological readjustments. Thus, with hot water extraction and cold water reinjection, fracturing and seismic activity may increase.

Geothermal Area Physical Properties

The temperatures of geothermal fluids in Imperial County KGRAs have been measured to range from a high of over 330°C to intermediate temperate systems of 90 to 150°C. The hottest area is the Salton Sea followed by Brawley, Heber, East Mesa, the border area, and the lowest in The Dunes. Apparently, these geothermal fluids will not flow from wells unless a borehole water column is stimulated by injecting nitrogen to the well bottom to lift the water column and allowing thermal expansion to occur. Once the flow is initiated, however, the fluid continues to flow without further aid.

A most important source of ground water in the Salton Sea area is the Colorado River. Apparently precipitation and local runoff contribute only a small fraction of the total storage in the Imperial Trough. Generally, there is a northward underflow from the Mexicali Valley, an underflow through the alluvial section between the Cargo Muchacho Mountains and Pilot Knob, a section above which the Colorado River meanders into the Gulf of California. Since the construction of the All-American Canal, and subsidiary irrigation canals, there has been some leakage from the canals and some leakage from overirrigation. The ground water table has risen 40 feet along the All-American Canal between 1939 and 1960. Much of the water leaked from canals drains into the Alamo and New Rivers and thus ultimately in the Salton Sea.

The origin of the salts of the Salton Sea geothermal brines is not clear. Some argue that the brine is from the flow of Colorado River water and others have suggested that it is formed by leeching of sediments by local precipitation circling downward into the geothermal reservoir. There is great variation in the salinity of the reservoirs in the Imperial Trough. Not only does the salinity composition vary by KGRA, it may even vary from well to well in a given geothermal field. Further, measurements made at different times from the same well show some minor variations. For a survey of the salinity chemical composition of the various geothermal fields, one should see studies cited in Biehler and Lee (1977).

Effects of Fluid Withdrawal and Reinjection

Since most reservoirs are large compared to the amount of fluid mass withdrawn or reinjected, pressure variation as a result of these processes is not very great. Apparently, fluid composition will not change very much if there is no recharge. On the other hand, if the geothermal reservoir is recharged artificially or naturally, chemical composition is expected to change, depending upon the extent of recharging. In any case, there will be an interaction between the withdrawal and reinjection of fluids. The introduction of alien water, unlike the geothermal fluids in the reservoir, will offset whatever rock–fluid equilibrium exists. Fluid composition cannot be estimated from a simple mixing model because of the lack of knowledge about the size of the geothermal reservoir.

Utilization of the geothermal resource by withdrawing fluids may induce earthquakes owing to thermal strain and release. Thus, seismic hazards and rates of land subsidence may be increased in the vicinity of geothermal development (Biehler and Lee, 1977). This is probably true even if extraction is compensated for by reinjection of natural recharge or of other fluids. This probably can be expected because there will be thermal contraction in a reservoir with decreasing temperatures. Some estimate may be gained from the likelihood of these events as a result of the microearthquake study of The Geysers located in northern California (Bufe and Lester, 1975). Generally, earthquakes or seismic activity near The Geysers' geothermal development were shallow and of the normal faulting type. Unfortunately, however, seismicity in The Geysers area was not known before commercial development of the resource; thus, there is no knowledge of how much of the activity was caused by geothermal energy extraction. On the other hand, since the Salton Sea geothermal resource has not yet been commercially developed, it will be possible to estimate earthquake results from geothermal resource utilization. Also, it is imperative to estimate the earthquake potential in advance because of the possibility of subsidence and its potential for problems related to agricultural production.

Earthquakes are associated with stress. The extraction of geothermal fluids may induce thermal stress adding to the already existing tectonic stress, perhaps then resulting in movement along already preexisting fractures. Biehler and Lee (1977), using a variety of assumptions about releasing thermal stresses, related energy extraction to the expected magnitude of earthquakes. Their general conclusion was that thermal strain is likely to result in some small increases in earthquake scale and frequency but without great environmental damage to Imperial County. Generally, there is a suggestion that the magnitude of events will be small but with relatively frequent occurrence. In estimating earthquake frequency and magnitude as a result of geothermal resource utilization, Biehler and Lee suggest that a 100 MW_e yr power production probably would result in an earthquake magnitude of 4.1 on the Richter Scale and for the 1000 MW_e yr. production, 4.8 on the Richter Scale with earthquake frequencies increasing from 2 to 6.5 times per year, respectively. On the other hand, natural cooling in the Salton Sea field at the rate of 8 MW_e may result in an event of a magnitude of 3.4 with approximate occurrence of three times per year.

Life Expectancy of the Geothermal Resource

Generally, the life expectancy of a geothermal resource depends primarily on the total heat in storage and the rate at which additional heat is supplied, and, of course, the production rate. The production rate can be influenced by resource utilization. Generally, life expectancy calculations for Imperial County suggest that the resource has a finite life span. This life span ordinarily is estimated to have a range of 150 to 500 years if 5,000 MW_e is produced every year and the effects of thermal recharge are neglected (Biehler and Lee, 1977). Generally, these estimates are crude because of the uncertainty of making *assumptions*. Biehler and Lee (1977) point out that the "development of new technology which would increase the efficiency of thermal conversion or the discovery of attitional geothermal reservoirs could easily double or triple these estimates."

Estimates of the MW_e potential in Imperial County have substantial variability. One estimate places the potential at 4000 MW_e for 30 years, while Ermak (1978) gives a range of 3000 to 5000 MW_e for a 30-yr period. Layton (1978) suggests between 4000 and 5500 MW_e potential for 25 years. Thus, a minimum estimate of 1600 MW_e potential for 30 years used by some seems very reasonable (see Table 2-4).

One issue involved in the utilization of known geothermal resources is ownership. Geothermal fluids consist of water, dissolved minerals and gases, and heat. In some places, pressure is also a feature of the resource. Surface water, generally, is accepted as being a public resource while minerals and gases are most often associated with the subsurface and may be severed from surface estate holdings. Pressure has not been previously appropriated, and as a result, there is no established legal ownership history. Geothermal resources, then, ensure that a variety of people, organizations, corporations, and governmental units at the federal, state, and local level will be involved in their development.

Also, land ownership may be in private hands, or owned by the federal, state, or local government; this is highly variable among the states with geothermal resources. In California, about 2.8 million acres, of which 529,000 acres are minerally reserved, belong to the state. Also in California, acreage with minerals reserved for the U.S. is well over 2,300,000 acres, with approximately 157,000 acres for oil and gas, and miscellaneous other areas allocated to the states. In most states, surface ownership patterns do not reflect necessarily established resource rights since in many instances these have been sold without the current landowner owning them. Ground water, in most states, is controlled by the state as a public resource. However, this is not the case in California, in which water rights are adjudicated. In any case, water ownership rights are a potential source of problems.

The geothermal resource in Imperial County is water dominated and at most places is located at a range of about 2,500–12,000 feet.

Thus, it is not completely clear whether subsurface mineral rights should hold or if "surface" rights is the more appropriate legal perspective.

Perhaps of more immediate importance, however, is the need for water for cooling towers, etc., so that the resource can be exploited. Currently, county officials have committed water only to the demonstration plant phase; the pressure for additional allocations of water will be much greater after successful demonstrations. Geothermal energy production requires more water than fossil fuel or nuclear production. At the present time, the Colorado River is one source of the large amounts of cooling water necessary for geothermal energy development on a large scale. Clearly, there is substantial potential conflict that may develop between agricultural interests, who now use this water, and energy developers.

It has been estimated that to sustain production of 5,500 MW_e of geothermal energy, more than 300,000 acre ft of fresh water will be required for the cooling towers alone, or about 10% of the water that is currently allocated annually to Imperial County users from the Colorado River (Layton, 1978). Obviously, this is an important problem and an area of potential conflict since all of this water currently is allocated to agricultural uses.

There are differing amounts of cooling water needed for varying power plant levels. Ermak (1978:214) argues that steam condensate is sufficient to meet cooling water demands with no additional water being required if reinjection is not required. However, this situation is quite different if total reinjection is required. There is clearly some question as to how much water will be required for cooling purposes in the geothermal development, especially at its maximum capacity. Ermak (1978:214) estimates that the cooling water in acre ft/yr may be as high as 600,000 at the 8,000-MW_e level. He argues the most promising source of cooling water is the agricultural waste water flowing into the New and Alamo Rivers which, he believes, is of sufficient quality to support over 8,000 MW_e of geothermal electric power. Nevertheless, chemical treatment will probably be required before it can be used for cooling purposes. Also, the flow of water in these rivers is not uniform and there are distribution and storage problems which will probably rise at the more intensive levels of geothermal development. Also, the loss of water to the Salton Sea through geothermal development usage could result in the Salton Sea decreasing its elevation.

Large quantities of water are needed for use in evaporative cooling towers in order to reject waste heat from geothermal power plants.

The total water requirement is the sum of evaporative losses in the cooling tower plus blowdown water. Evaporative losses for the operating power plant in cooling water blowdowns suggest that at least 60 acre ft of water will be needed per MW_e in the Salton Sea KGRA and 90 acre ft in all the other Imperial County KGRAs (Ermak, 1978:211–212). The reason for the lower requirement for the Salton Sea is that, owing to differences in the geothermal brines, engineering design for Salton Sea power plants will incorporate an average energy conversion efficiency of 0.14, while designs for the other KGRAs will utilize conversion efficiencies of 0.10.

The water resources potentially available to meet these requirements are Colorado River water, agricultural waste waters, the Salton Sea, ground water, and condensate from geothermal steam (Layton, 1978:133). A small additional source is municipal waste water. However, there are several dilemmas and difficulties in acquiring adequate sources of cooling water and there are also the consequences of using a particular water supply. Virtually all of the water resources in Imperial County are highly saline, with the Colorado River water containing nearly 1.2 tons of salt/af of water or about 882 ppm total dissolved solids (TDS). Waste water from the agricultural drainage system has even greater solid waste and salt concentrations. Much of the brackish waste waters from both the Coachella and Imperial Valleys end up in the Salton Sea. When equilibrium exists between the inflows and evaporation of the Sea, the salinity of the Sea increases steadily and now is approaching 40,000 ppm TDS. In recent years, the influx of water to the Sea has been greater than the evaporation rate, thus producing increases in the Sea surface elevation and overflowing the former shoreline in certain areas.

Ground water contained in sediments underlying the valley is another resource; however, it also has a high salinity level. The most promising area for ground water extraction is in the East Mesa region, where recharge from the unlined All-American and Coachella Canals has amounted to well over 7,000,000 af since 1950 (Layton, 1978:137); other ground water estimates discussed in Chapter 8 are higher. In addition to Colorado River water, agricultural waste water, and ground water, geothermal steam condensate can be used for cooling. Municipal waste water constitutes a small additional source.

There are constraints on use of the various water sources in Imperial County. For example, imported Colorado River water is the most desirable water for use in power plant cooling systems but the water is distributed in the valley currently to support irrigated agriculture and it is unlikely that substantial quantities of irrigation water will ever be available for nonagricultural uses. The brackish waste water contained in the New and Alamo Rivers and in the drainage structures of the irrigation district is another supply possibility. The dilemma in using this water involves the disposal of saline blowdown waters resulting from the use of brackish waters in cooling towers. Since much of the water in Imperial County already is highly saline, disposal of this water into drainage ditches would not be allowed under present regulations since it would further decrease the water quality of the region. One possible solution is to dispose of the waste in lined evaporation ponds, but Imperial County has restricted the amount of land that may be used for geothermal projects. Further, salt stored in the ponds would ultimately have to be removed to a Class I waste disposal site, and none currently exists in the entire region (Layton, 1978).†

Before some of the geothermal condensate can be used, there will have to be studies on the amount of subsidence that occurs in the county as a result of geothermal development:

> If increased subsidence does occur with geothermal projects, new policies may be established that require complete reinjection of the geothermal fluids extracted from the reservoir. A policy of total reinjection would mean that additional water supplies would have to be acquired to replace the condensate used to make up water in cooling towers operated with flash steam systems (Layton, 1978:141).

If water is used from the Salton Sea, a waste water diversion of 60,000 af/year to support a generation capacity of 1,000 MW_e would eventually result in a decline of 2½ ft in the Sea's elevation and a salinity rise of 3,500 ppm greater than the currently expected value. So the utilization of water flowing both into the Sea and from the Salton Sea itself poses certain kinds of problems. These may appear to be moot at the moment since the Sea is rising each year rather than declining. Nevertheless, the longer-term view suggests that the use of this water could result in a decline in the Sea's elevation and an increase in salinity, which is now bordering on eliminating all wildlife from it.

A number of constraints regulate power plant siting. At the federal level, the Secretary of the Interior has the responsibility for designating endangered and threatened species, critical habitats, and wildlife refuges and for leasing geothermal resources on federal lands. Other license-granting authorities are generally held by state agencies,

---

†California has recently established a Class I waste site for all geothermal solid waste from Imperial Valley.

e.g., leasing state lands. Similarly, local governments exercise control over power plant siting by regulating permits, zoning ordinances, and by the authority to protect local citizens' health and welfare.

The Secretary of Interior and California Department of Fish and Game have listed five endangered species in Imperial County. This also will affect where geothermal power plants can be sited. The California Energy Resources Conservation and Development Commission also has the state licensing capability for power plants over 50 MW_e. Thus, some plants will probably generate only 48 MW_e to avoid one more regulating agency!

Another area of concern in the development of geothermal resources in Imperial County is land use. The land surface area of the well field required to support a 100-MW_e power plant for 30 years varies with characteristics of the geothermal resource. Ermak (1978) gives a general estimate of approximately one square mile plus to support a 100-MW_e power plant. However, the major factor controlling well field size is the temperature of the geothermal fluids, with higher temperatures requiring fewer wells and, therefore, less land area. Results in Imperial County suggest that the size of the well field in the Salton Sea KGRA will be approximately half as large as those in the other Imperial County KGRAs for the identical capacity power plant. This is primarily because the Salton Sea fluids are much hotter than those located elsewhere in Imperial County. Thus, the well field size required to support a 100-MW_e power plant is 1.25 square miles for the Salton Sea KGRA and 2.50 square miles for Brawley, Heber, and East Mesa KGRAs (Ermak, 1978:207).

As we have already noted several times, most of the area in which geothermal plants can be located is on agricultural land and Imperial County is one of the most valuable agricultural resources in the nation. The Geothermal Element (1977) of the county plan and official county policy is to minimize the land area occupied by geothermal facilities, including wells, pipelines, generator buildings, storage areas, and so forth.

A total of 35 wells will have to be drilled to support a single 100-MW_e power plant in the Salton Sea KGRA for 30 years. Power plants located in other Imperial County KGRAs will require as much as twice the geothermal fluid to produce the same electrical output owing to the lower temperature of the fluids; thus, these will require approximately 70 wells to be drilled for each 100-MW_e power plant.

At the highest level of expected geothermal power development of 8000 MW_e, less than ½% of the existing agricultural land area is expected to be used. Thus, the amount of land area that will be used in direct geothermal facilities and energy production is relatively small. However, not systematically addressed yet in the county is how much land will be lost indirectly to geothermal activities; this amount of land lost can be significant if not properly planned for. At higher levels of development, power plants probably will spread over large areas of agricultural land with pipelines running from wells to the power plants, transmission lines carrying the electricity from the power plants to users, and canals bringing cooling water to the power plants. All of these could possibly disrupt current agricultural practices.

All of these comments are related to electrical generation. There are, of course, two possibilities in utilizing geothermal resources. The first is to generate electricity and the second is for nonelectrical uses. Both electrical and nonelectrical use require plants within several miles of the resource. Although, technically, brines can be transported up to as much as 50 miles, the economics of transmission pipes (see Chapter 8) limits transmission distances generally to several miles. An electrical generating power plant must be located near the resource since geothermal fluids cannot be transported over long distance to a power plant that may be located at a more favorable position; the energy must be converted to electricity in the general vicinity of the well field (Ermak, 1978:206). Similarly, nonelectrical use must be located near the resource or the fluids will not be usable.

Conclusion

This chapter has looked at the origin, extent, and utilization of geothermal resources—in Imperial County and worldwide. Although potential worldwide resources are vast, many world regions endowed with geothermal resources are hampered by regulations and/or lack of interest and investment capital. Currently in the U.S., the interest of government, private enterprise, and the general public is focused on energy problems. Why, then, have not the large U.S. geothermal deposits, some discussed in this chapter, been more swiftly developed? Among other reasons, Federal, state, and local regulations have delayed and, in some cases, prevented development. Some regulations are geared to other types of resources and need to be streamlined or revamped for geothermal purposes.

Worldwide, geothermal development will undoubtedly achieve much greater importance by the year 2000. The year 2000 estimate of 75,000–100,000 MW of generating capacity does not include the potentially larger energy expenditures in direct uses of the energy. Since some of the largest utilizations are projected to occur in the advanced industrial nations of Japan and the U.S., technological advances may be expected to occur beyond the present geothermal technology outlined in this chapter.

ACKNOWLEDGMENT

Parts of the section on Geothermal Resources in Imperial County were drawn from the report by Shawn Biehler and Tien Lee titled "Final Report on a Resource Assessment of the Imperial Valley," NSF/ERDA grant No. AER 75-08793, Dry Lands Institute, University of California, Riverside, 1977.

3

Population–Economic Data Analyses Relative to Geothermal Fields, Imperial County, California

INTRODUCTION

Geothermal energy worldwide exists in diverse geological, ecological, climatic, and socioeconomic settings. The oldest developed field in Lardarello, Italy has been producing electrical energy since 1904 (Wehlage, 1975). Lardarello has geologic features of steam fields, large production area, and others. Current production is 404.6 MW$_e$ (Wehlage, 1975). The socioeconhomic setting has included much overlapping agricultural and geothermal use, rural property in surrounding regions, relatively small amounts of nonelectrical use, long distance from large Italian population centers, and since 1967 a national policy of control of resource development by the Italian Federal Energy Agency.

In Sonoma County, California, the single U.S. field of substantial electrical development (over 600 MW$_e$ in 1979), a rich steam resource underlies a hard rock geologic strata. In its early history this resource manifested itself in vacation spas with hot baths, and many hot springs. From a socioeconomic standpoint the Sonoma–Lake County resource is located underneath a sparsely populated, hilly area containing retirement and vacation areas and a few small and rather unproductive farms. Vollintine and Weres (1976) estimated a population of 50 persons in the field location of extreme northwest Sonoma County, and 5000 persons who might be exposed to geothermal development in Lake County.

Although there are few people in surrounding areas and nearly all of the energy is placed on power grids to northern California population centers, the local population and economic characteristics are important to energy development. From an energy development standpoint, they act as geothermal labor supply (more highly skilled labor is mostly brought in from outside), sellers and renters of geothermal land, and as a supplier of indirect (and some direct) services to support geothermal operations. From a broader perspective, they are important as voters and contributors to public opinion on geothermal issues—which influences public policy. The importance of such public opinion is evident in the environmental restrictions on geothermal development in Lake County (Vollintine and Weres, 1976).

Besides Lardarello and Sonoma County, there are other examples of adaptations of human communities to geothermal resources. This chapter examines the historical and current population characteristics of Imperial County, California. These include such features as vital rates, urbanization, town sizes, labor force composition, income, utility usage, and ethnic composition. From these basic data, inferences are drawn on some of the important social and economic processes related to geothermal development. Multivariate statistical analysis (discriminant analysis) is used to study relationships among variables. Another intent is to speculate on the adaptation or range of adaptations that the county's population will follow, as geothermal plant capacity is scaled up. Such speculation may only be confirmed by standardized comparisons after an energy capacity has been added.

To accomplish the above goals, this chapter is divided into several sections. One section reviews historical studies and past research which predates in methods or results to the present work. Another section analyzes population trends for the county as a whole. Since U.S. Census data on counties were collected in much greater detail beginning with the 1950 census, nearly all data series and trends are for the period 1950–1970.

Subcounty data are analyzed for 1970. The geographical unit used here is the enumeration district, a U.S. Census population grouping with an average size in Imperial County of 931 persons. Regional trends in characteristics such as ethnic composition, dependency ratio, and mobility are studied, and compared with each other for energy development implications. This latter analysis is carried further in a section investigating characteristics distinguishing enumeration districts overlying geothermal fields from those that do not. Since ethnicity is of key importance in Imperial County, one section is devoted to examining the county's largest ethnic group—Spanish-Americans.

Energy derived from nuclear, fossil fuel, coal, and geothermal sources may be placed on an electric power grid, and for small incremental costs, conveyed to localities hundreds of miles away (Fowler, 1975). Such power grid relationships vary according to season, regional supplies, alternate forms of energy, utility regulating decisions, international pricing developments, and so on. Regional use of geothermal power has the complication of changeable and distant dispersions by power grids, as well as plant capacity.

Regional relationships may be assessed, however, for nonelectrical uses of geothermal power. Geothermal energy production is based on subsurface hot water or steam, which in turn is heated by elevated magma layers. In geothermal power development, wells are drilled into the geologic strata: hot water or steam is withdrawn and used to produce steam to turn generators which yield electricity. The regionally fixed product of hot water may be used within about 60 km of the power plant for heating, air conditioning, industrial plant processing, or for agricultural purposes, and recreational use. Both electrical and nonelectrical uses have population–economic implications for local areas.

This chapter analyzes the socioeconomic characteristics of Imperial County, California, in relation to known locations of geothermal fields, which are identified as KGRAs. It should be noted that KGRAs are areas for which the federal government has estimated potential geothermal production. These estimates are often old and often do not accurately portray the extent of potential commercial production. Socioeconomic data are from the 1970 U.S. Census. Analytical methods we use are graphic displays, demographic techniques, and statistical discriminant analyses.

There have been few studies of socioeconomic effects of geothermal energy development. A recent study examined public opinion on geothermal energy development in Lake County, California, the major existing site of geothermal energy production in the U.S. (Vollintine and Weres, 1976). Generally, population variables were concluded to be unrelated to opinion on geothermal energy development. Kjos (1974) investigated the potential for industrialization in Calexico, Imperial County's sister of Mexicali. He drew essentially negative conclusions about industrial development of Calexico due to the lack of incentives for new businesses, the lack of governmental planning, local tensions in regard to border worker issues, and a lack of an industrial park, among other reasons.

POPULATION ANALYSIS OF IMPERIAL COUNTY

Imperial County is located in the southeast corner of California; its central valley portion, geologically, is part of the Salton Trough.

Before 1901, the valley consisted of Sonoran Desert, with only several thousand residents in a land area of 4284 square miles. The 1901 diversion of Colorado River water for irrigation was accompanied by an increase in population. Several years later, the northern drainage flow of an accidental flood of irrigation water and drainage from the irrigation process resulted in the Salton Sea, an undrained salt sink.

Generally Imperial County has not followed the urbanization–industrialization trend characterizing Calfornia's population history (Thompson, 1955). In the U.S. Census of 1970, Imperial County was 32.1% rural, compared to 9.1% rural for the state. The state, by comparison, reached the 32.1% value in 1920. Small business size (less than 60 employees) characterizes the Valley's business structure (Imperial County, 1973). Such a rural, small business, small town character is a rare feature in contemporary California.

The Valley also differs from statewide history in population size trends as well. Subsequent to the onset of irrigation, the population rapidly increased to 60,930 persons by 1930. At that time the population reached a plateau; in the next 40 years the population increased only 22%, compared to 251% for California. Even including, in 1970, the 9,000 estimated Mexican border commuters, population only increased 37%. A trend which paralleled total numbers was a decrease from a 1910 sex ratio (ratio of males to females) of 1.9, 52% higher than California, to a steady level 18% to 26% higher than the state's from 1930–1960. If international commuters (nearly all males) are included, the relative sex ratio continued 28% higher than the state's for 1970.

Since agriculture has consistently been the base industry of the county, the fundamental economic support for additional persons has not been significantly altered since 1930. The attraction of large amounts of developing industry and commerce, which altered other formerly agricultural counties like Orange in this period, was probably precluded by labor and locational factors. The county's unusual sex ratio is interpreted as a lack of female occupations and amenities during the pioneering years of 1910–1930, followed by an employment need for more males in agriculture since 1930. Total populations of state and county for 1910–1970 are given in Table 3–1.

The wide growth differential is not accounted for by differences in fertility or mortality, but by differences in interstate and intrastate migration patterns. Throughout urban California, industrial growth has continued to attract migrants from the East. However, a well-known feature of the agricultural component of the U.S. has been a trend towards increasing mechanization and decreasing manpower re-

TABLE 3-1. POPULATION AND SEX RATIOS FOR CALIFORNIA AND IMPERIAL COUNTY FOR 1910 TO 1970[a]

Year	Imperial	SR1	California	SR2	SR1/SR2	Mexico
1910	13,591	1.90	2,377,549	1.25	1.52	15,160,369
1920	43,453	1.48	3,426,861	1.12	1.32	14,334,780
1930	60,903	1.36	5,677,251	1.08	1.26	16,552,722
1940	59,740	1.23	6,907,387	1.04	1.18	19,653,552
1950	62,975	1.19	10,586,223	1.00	1.19	25,791,017
1960	72,105	1.24	15,720,860	0.99	1.25	34,923,129
1970	74,492	0.98	19,953,134	0.97	1.28	48,313,438
1970[b]	83,492	1.24	—	—	—	—

[a] SR1, Imperial County sex ratio; SR2, California sex ratio. Source: U.S. Bureau of the Census (1910–1970).
[b] Adjusted to include border commuters.

quirements. Hence, growth of agriculture in Imperial County has not spurred a major population increase over the past 45 years.

In fact, owing to natural increase, the above figures indicate a net migration loss to the county in every intercensal period since 1930. Generally, it has been shown in population research that an economic, job-seeking motivation is the primary determinant of migration in the U.S. (Saben, 1964; Lansing and Mueller, 1967; Morrison, 1972). Thus, one can speculate on greater job attractiveness pulling some persons (mainly young adults) away from the county.

The most recent state projection figure for the county in 1975 is 84,100 (State of California, 1976). If the projection is accurate, it would indicate at least a net migration which is balanced to zero—a change in prior trends. This would correlate well with U.S. Census findings of the past several years, showing a nationwide trend towards return to rural areas. Geothermal development will either be negligible in effect or increase in-migration. Projection of future county population (see Chapter 5) features migration as the critical component.

Several other general features of the county should be noted. As shown in Table 3-2, in 1970 the county was 46% Spanish-American, according to the census designations of either Spanish language or Spanish surname, also called Spanish "heritage." This compares to 14% for the state. Spanish language is estimated postenumeration by responses to the census question, "What language other than English was spoken in this person's home when he was a child?" Spanish language means postenumeration expansion of the Spanish mother

TABLE 3–2. PERCENT SPANISH-SPEAKING POPULATION IN IMPERIAL COUNTY 1930–1970[a]

Spanish-speaking definition	Percent of population		
	Imperial	Kern	California
1930: Mexican	35.5	8.6	6.5
1950: Mexican birth	13.5	1.5	1.5
Mexican origin (est.)	37.1	4.1	4.1
1960: Mexican origin	27.2	5.0	4.4
1970: Mexican birth			2.0
Mexican origin	30.4	7.1	5.5
Spanish language	44.9	15.5	13.7
Spanish language or Spanish surname	46.0	16.8	15.5
Spanish origin or descent	40.3	15.1	11.8
Spanish mother tongue	41.3	13.0	10.7

[a] Mexican origin for 1950 estimated from 1970 data. Source: U.S. Bureau of the Census (1950–1970).

tongue designation to include all persons in households with Spanish mother tongue for the head or spouse (Hernandez et al., 1973).

The 1970 Census Spanish surname designation is established by comparing a list of 8000 Spanish surnames with respondent names on 20% of 1970 Census questionnaires in the Southwest. The designation Spanish language or surname is thus quite broad. Another designation (30.4% for the county) is Mexican foreign stock or Mexican origin, defined as birth in Mexico or in the U.S., for which one or both parents were born in Mexico. The definition of Spanish language alone gives a percentage of 44.9%. Finally, Spanish mother tongue consisted of 41.3% of Imperial County's population.

Demography of Spanish-Americans has been researched extensively (Grebler et al., 1970; Bradshaw, 1973; Uhlenberg, 1973). A key conclusion of these studies is higher fertility for Spanish-Americans (henceforth, Spanish-American is abbreviated as SA) compared to the anglo population. In the 1960 Census, SA fertility, measured by total childbearing per woman, was 80% higher than that of anglos. For 1970 Census data of five southwestern states including California, Bradshaw (1973) gives a total childbearing rate which is elevated 61.1% for persons of Mexican origin over anglos, and 45.1% higher for Spanish language or surname persons over anglos. In Imperial County, this influence is revealed by the county's leadership in fertility among all California counties (State of California, Department of Public Health,

1970). Even assuming a 50% average extra fertility per ten-year intercensal period, ethnicity would account only for about 5% addition of population over ten years, a small increment compared to in-migration which characterized the state from 1930 to 1970.

Other population characteristics of the county are low average income, urban concentration in towns of 4,000 to 20,000, low average educational attainment, low average age, and occupational concentration in agriculture and related areas (Imperial County, 1973; U.S. Census, 1973). A special occupational feature of the county is the proximity of the international border. A labor pool, with residence in Mexicali, travels to work in the county every day, and returns in the evening. This pool has important seasonal fluctuations (U.S. Senate, 1970; Kjos, 1974; Nathan Associates, 1968), with peak daily border crossings in January and late May, in correspondence to crop harvesting demand. In the early 1960s, daily commuters varied seasonally from 4700 to 8000 per day (U.S. Senate, 1971). There are also legal and illegal aliens residing in the county, many of whom add to the labor force (U.S. Senate, 1971).

Because of the large labor pool originating in Mexicali, it is important to look at the population growth south of the border, 1950–1970. Table 3–3 shows the growth of Mexico and Northern Baja California for these years.

In contrast to Imperial County, the population size of Baja has more than doubled for the last two 10-yr periods, whereas the population of Mexico has increased by approximately 40% each period. North of the border, population is lagging behind statewide and national trends; south of the border, the Mexicali region and Northern Baja are greatly exceeding national trends. One can easily surmise here a potential pressure for increased legal and illegal border crossings. Mexicali itself has increased greatly to become the fifth largest Mexican

TABLE 3–3. POPULATION GROWTH FOR NORTHERN BAJA AND MEXICO[a,b]

| | | | Mexico | |
Year	Northern Baja	Mexico	Birth rate (%)	Death rate (%)
1950	227	25,791	45.5	16.2
1960	520	34,923	44.6	11.2
1970	1,105 (est.)	48,313	46.0 (1975)	8.0 (1975)

[a] Modified from Whetten (1971).
[b] Population is in thousands.

Figure 3-1. Downtown Heber.

city, rapidly approaching the half-million size (Kjos, 1974). Mexicali's size in the 1980 Mexican census is likely to exceed 700,000. However, since the number of "green card" holders is, in the final analysis, a diplomatic decision of the U.S. Federal Government, border crossings have remained steady. Illegal immigration is postulated to have increased greatly in the last ten years (Dagodog, 1975).

A fundamental factor in Imperial County's population, then, is the Mexican border's large number of Spanish-Americans (SAs). Although sparsely populated in the early 1900s, the northern region of Mexico has increased in population sharply since 1950 to reach a 1970 level of 857,000 people with 390,000 residing in Mexicali, immediately adjacent to Imperial County. Table 3–2 shows comparative data on the SA population 1930–1970. Comparison of the broadest definitions, "Mexican" for 1930 and "Spanish Language" for 1970 reveals substantial equilibrium between 1930 and 1970 for the county. Imperial County's fivefold larger SA percentage than the state's in 1930 was likely due to border proximity and to Mexican emmigration (Samora, 1971). Areas near the border, such as Heber, Figure 3–1, house a higher proportion of SA population.

The stabilized nature of the county's agricultural economy, set in

1930, has not offered an employment structure able to accommodate a skilled mix of increased SA workers since that time. Noncitizen, Mexican-origin residents have been present during this history. There are legal Mexican aliens (LMAs) and illegal Mexican aliens (IMAs). The detailed history of LMA and IMA trends and fluxes since 1900 has been delineated by Samora (1971). LMA numbers have been influenced by economic demand for cheap labor, resulting in a federally approved contract labor program in the 1950s—the Bracero Program. This program peaked with 400,000 contract workers in the U.S. during the late 1950s. Since U.S. place of residence alone determines the census, substantial numbers of braceros were counted in the 1960 County Census. With the ending of this program in the 1960s, some 4,500 to 6,000 farm workers established residence in Mexicali and commuted to jobs in the county daily (U.S. Senate, 1971). The daily influx by Mexican workers to the U.S. is estimated to number presently between 6,000 and 12,000. The Immigration and Naturalization Service estimated county LMAs at 8,000 in 1973. Recent literature has pointed out the extreme inaccuracy in estimates of IMAs (Stoddard, 1976).

The above picture of Imperial as an agricultural, slow-growth region may be altered by the advent of geothermal resource development also many people do live in towns such as Brawley (Figure 3–2).

FIGURE 3-2. Town of Brawley, Imperial County, California.

For population projections, it is important to speculate on possible change in the demographic determinants. Mortality levels are not postulated to be lowered significantly in the near future. Although fertility is likely to continue to be amplified by the large number of SAs and to be cyclical in accordance with past U.S. patterns, geothermal energy development would appear to have possible fertility influence only through the mechanism of reducing the SA percentage of the population through an in-migration increment of skilled workers, with a smaller percentage SA.

MIGRATION

Significant geothermal population change may take place in the future as a result of geothermal development if in-migrants are attracted. The key factor appears to be potential industrialization of the county related to geothermal energy. Before examining causes of industrialization more specifically related to Imperial Valley and geothermal energy, it is important to look at general studies of industrialization in rural areas.

Ginther et al. (1974) studied factors in location of 330 out of 1900 present industrial plants in Nebraska. The following question was asked: "At the time the decision was made, how important were the following factors in selecting your present location?" Responses were graded on a scale 0–3, and revealed the following seven most important factors out of 43:

1. Labor quality
2. Highway transportation
3. Labor availability
4. Available site
5. Reliability of electric service
6. Wage rates
7. Proximity to market

Electric rates were 12, and eight community attractiveness factors were rated between 17 and 35. Although energy was of some importance, the major decision influences were labor force and transportation. Since neither in quality nor quantity of labor force does Imperial County compare to such metropolitan regions as Los Angeles and San Diego, this is a somewhat pessimistic reflection on geothermal industrial potential, at least the electrical component, in Imperial County.

A second study of rural industrialization, written from the experienced viewpoint of an executive in the consulting business of locating plants, noted a general trend of locating new plant locations in rural areas (Fulton, 1974). For instance, one-third of new manufacturing job openings in 1962–1969 were in small, nonmetropolitan areas. Causes alluded to were journey to work frustrations, fear of physical attack, air and noise pollution, high living costs, and economic anxiety—50% higher living cost in New York City than in communities of less than 50,000. With the large suburban growth of 1960–1970 (Hauser, 1971), why did industry not choose mainly suburban locations? Fulton notes suburban problems of pollution, blight, bad land use, transportation difficulties, intergovernmental conflict, rising land costs, and labor shortages. A rural location was favored by closeness of workers to the job, more space for dispersion of plant pollutants and waste heat, cheaper and more available land, greater governmental interest, etc. The best communities for rural industrialization had the following: good highway system (preferably a national interstate), a strong leadership in the community who are informed about government programs, nondomination by a single large industry influencing the labor market, labor availability, and easy access to the facilities and services of a large city (less than 50 miles distance). The conclusions for Imperial County to be drawn from the above studies are twofold: first, the recent national trend is most favorable to not-too-isolated semirural areas; and second, geothermal electricity may be secondary rather than primary as a factor in decision-making on industrial plant locations. Geothermal industrialization of nonelectrical type may offer greater reason for choice of location in the county.

Kjos (1974) investigated the potential for industrialization specifically in Calexico, the sister city of Mexicali. His study is historically illuminating, as well as pinpointing some of Imperial's issues on entry of new business.

Kjos traces the history of labor interaction between the sister cities. The Bracero Program of 1951–1964 involved contracting of Mexican labor to work in the county. When this program ended, the Mexican government gave increasing emphasis to building Mexicali's economic attractiveness. An outgrowth of this was the Border Industrialization Program, initiated in 1971, which encouraged mutually beneficial industrial development on both sides of the border. Kjos regards this program on the Mexicali–Calexico area as essentially unsuccessful. Thus, his study asks the question, "What does Calexico wish to do to encourage its own industrial buildup?" He interviewed Calexico citizens, Mexicali business and government leaders, American businessmen who considered locating in Calexico, and American businessmen participating in the Border Industrialization Program. His conclusions are essentially negative about impetus for industrial development of Calexico, owing to lack of incentives for new businesses, lack of governmental planning focus in Calexico, local tensions in regard to the border worker issue, lack of an industrial park, and other reasons. In general, he documented community indecisiveness regarding industrialization. It is difficult to apply results of his study outside of the very unique border situation, but possibly community decisiveness will be of importance to geothermal industrial development.

In the long run, Imperial County's agricultural nature may be advantageous for geothermal energy development. It is important to note that agriculture and food processing are estimated to consume about 2.5% of the total energy consumption in the U.S. (Steinhart and Steinhart, 1974). Thus, the very large agricultural component of the county economy (Force, 1970) may expand further with larger energy sources. Increasing inputs of energy into one crop, corn, tripled from 1945 to 1970 (Pimental et al., 1975).

Perhaps the most difficult technological part to project in a geothermal scenario is nonelectrical use of potential production. Possible uses include heating, air conditioning, production of chemicals from brines, heat for industrial processes, such as food processing, desalting of salt water, drying or distillation of every type, heating of soil, raising of fish, etc. (Lindal, 1973; Wehlage, 1974; Barnea, 1972; James, 1970; Werner, 1970). An important consideration of such applications is that, unlike electrical use, they must be located somewhat near production sites. Thus, in all cases, this usage would draw industry into the county, rather than to external locations.

Energy Capacity and Consumption in Imperial County

The steady level of total population of the county is presently supported by an ample local energy capability, which evolved beginning in the mid-1930s. Since the first energy installation in 1936 of a 2.2-MW_e diesel generator, installed capacity increased nearly linearly (at a smoothed linear rate of 3.4 MW_e/yr) to 47.9 MW_e in 1950. From 1950 to 1975, power plant installation again increased about linearly (at a smoothed linear rate of 12.0 MW_e/yr) to 347.3 MW_e capacity in 1975. From 1940 to 1970, U.S. electrical energy production increased by a factor of 8.8 and per capita electrical production increased by 5.3 times (Fowler, 1975). By comparison, these factors for Imperial County energy capacity are, respectively, 12.9 and 10.4. Thus, while the county's population growth has lagged, its growth in energy capacity has greatly exceeded national rates. In fuel mix, the county has changed from steam–diesel–hydro proportions of 0:100:0 in 1940 to 52:7:41 in 1960 and to 68:11:21 in 1975. This latter compares to an estimated 1975 national mix of 76% petroleum and gas, 4% hydro, and 20% in other forms of energy (Penner and Icerman, 1974). Thus, the most unusual present feature of the county fuel mix is an enlarged component of hydroelectric power.

Per capita energy consumption was estimated by discounting national consumption in 1973 and county consumption in 1975 to 1970 by the national energy consumption growth rate of 5.6% per year (Fowler, 1975). Per capita estimates are 8272 kW hr for the nation and 7325 kW hr for the county (6530 kW hr if border commuters are included).

Lower comparative per capita consumption contrasts with higher installed capacity per capita. Whereas the 1970 total and peak demand capacities are 1.67 kW/capita and 1.35 kW/capita for the U.S. (Fowler, 1975), they are 3.59 kW/capita and 1.89 kW/capita for the county (3.20 and 1.69 with border commuters.) A final crude energy consideration is the current use of county electrical generation. The 1975 ratios of residential to commercial to industrial are 42:39:14 compared to 1973 national figures of 34:24:41. The housing fuel mix of the county reveals a major difference from the statewise pattern—reduced use of utility gas (37% below the state level) and increased use of electricity (271% above the state). A possible explanation is the low price of electricity due to the relatively large component of hydroelectricity in the county.

POPULATION COMPOSITION AND CHANGE

The population of Imperial County increased 22% from 1930 to 1970, a period in which the population of California increased by 251% (Table 3–1). Assuming an average rate of natural increase (the ratio of births minus deaths to total population) of 1.65%, population would naturally have increased by 66% with no net migration. Thus, for 1930–1970, there has been approximately a 0.7% net departure from the county per year. The most recent state projections show a July 1, 1975, population of 84,100. This projection assumes a yearly in-migration rate of 1.1% between 1970 and 1975, or a reversal of the 40-yr trend. This reversal, *if it has taken place,* corresponds to a national trend of return to rural places, which has been noted over the past several years by the U.S. Census. The state assumes about a 0.34% in-migration rate for California over the 1970–1975 period. These county projections do not employ a count of persons in the county, the most accurate method of estimating migration and one which will not be available until the time of the 1980 census results.

Although migration is the dominant influence on California counties, in Imperial it is reduced to the order of magnitude of vital rates—fertility and mortality. For the period 1970–1974, the county's average rate of natural increase was 1.23% compared to a state rate of 0.75%. The corresponding values for 1960 were 1.92% and 1.51%, and for 1950 were 2.28% and 1.38%. A standard demographic feature of SA population is higher fertility (Bradshaw et al., 1973). However, while the county's annual rate of natural increase is about 65% higher than the state's, this differential is only partly due to SA standardized fertility. For completed cohort fertility (i.e., total childbearing) in 1970 for ever-married women, ages 35–39, Imperial has a 49% additional increment of SA over non-SA fertility. When this is applied to the entire population, the fertility increment due to SAs is only 23%. The 42% differential is readily explained by the young age distribution of the county, especially of the SA population, since youthful age distributions tend to inflate standardized fertility and deflate standardized mortality.

Without this added fertility–age structure effect, Imperial would have declined in population by about 5% from 1930 to 1970. Several investigators have noted a convergence between Spanish-American and white fertility over the last 15 years, but such convergence does not appear in Imperial County. Perhaps this is because of the proximity to the border, the arrival of new Mexican migrants, and possible reduced acculturation and assimilation of the local Spanish-speaking population compared to the state.

To assess this fertility influence in the future, it is necessary to look at the census measurement of Spanish population for the last three censuses (Table 3–2). Five different classifications of Spanish are shown. This confusion of definitions was carefully clarified by Hernandez et al. (1973). A longitudinal perspective may be gained only through use of the definition "Mexican country of origin," which means that either a person was born in Mexico or one of his or her parents was. The 1950 definition of Mexican birth had to be converted to the above definition by an estimation procedure.

Although the percent of Mexican origin increased in California by 35%, 1950–1970, it declined by 18% in Imperial. This figure for Imperial is not strictly comparable, since many persons of Mexican origin, who might have been classified in the county in 1960, were living in

Mexicali and commuting in 1970. A rather stable level of SA population would appear correct for 1950–1970. This inference is further supported by low in-migration rates from other counties for Spanish-American population compared to whites (see later section on migration).

The potential effect of geothermal energy on the SA population may be a reduction in its proportionate size if geothermal energy brings large numbers of new persons into the county. Even if geothermal-related industrial demands are ideally suitable for this population, Federal immigration restraints will likely prevent the SA population from increasing, unless there is future change in border regulations.

Although of less importance for future demographic change in the county, mortality tends to reflect a general level of health and medical care. As shown in Table 3–4, Imperial had life expectancy in 1970 of 66.3 years for males and 75.2 years for females, values closely approximating state life expectancies. The crude death rates are lower than the national and California averages because of the young age distribution in the county. A reverse effect takes place with age standardized fertility indices, where the surplus of young persons increases the comparative differences with crude rates.

Age structures for the county from 1950 to 1970 are shown in Figures 3–3 through 3–6. Figure 3–3 shows the percentage change in the observed population. The noticeable drop in percent of males aged 20 to 39 in 1970 was partly caused by the change in status of bracero population residing in the county in 1960 to green-card commuter status with Mexican residence in 1970. A second likely data collection anomaly is the increased number of males 75+ in 1960. The figure shows large percentage alterations for female age groups 0–29 and 55+, and for male age groups 0–39 and 65+. The female side of this figure can be presumed to be less subject to SA work force factors for ages 20–64. Two trends are apparent: the presence of a depression-age gap in number caused by the maturing of cohorts of persons born in the low fertility years of the 1930s; and an increased percentage of females 30–39 in 1950, relative to 1960 and 1970, an effect possibly due to the large out-migration of younger persons of the 1950s and 1960s. The elderly (65+) population of the county increased by a factor of 2.5 from 1950 to 1970, and can be expected to continue increasing in the future. Another observation from this figure is the decrease in the sex ratio in 1970 as compared to 1960 and 1950. The decreasing trend in the sex ratio is apparent in Table 3–1 from the high levels early in the century. However, because of the presence of the green-card com-

TABLE 3-4. MEASURES OF FERTILITY AND MORTALITY FOR IMPERIAL COUNTY FROM 1950 TO 1974

Year	Births (Imperial)	Birth rate[a]		Total fertility rate[b]		Deaths (Imperial)	Death rate[c]		Life expectancy[a]			
									Imperial		California	
		Imperial	California	Imperial	California		Imperial	California	Male	Female	Male	Female
1950	1878	29.8	23.1	3836	3007	446	7.0	9.3	62.8	69.0		
1951	1990					445						
1952	2010					512						
1953	2301					475						
1954	2206					515						
1955	2184		24.1			463		8.8			67.3	73.7
1956	2146					489						
1957	2100					520						
1958	2003					486						
1959	1922					524						
1960	1885	26.1	23.7	4347	3622	498	6.9	8.6	74.2	73.4		

Year	Total births	Birth rate[a]		Total fertility rate[b]	Total deaths	Death rate[c]		Life expectancy[d]	
1961	2057				487				
1962	1959				526				
1963	1965				561				
1964	1882				591				
1965	1722		19.2	3211	587	8.3		67.8	75.1
1966	1755				557				
1967	1683				530				
1968	1613				590				
1969	1589				574				
1970	1662	22.3	18.2	2352	582	8.3	7.8	66.3	75.2
1971	1570	20.9	16.3		565	8.3	7.5		
1972	1444	19.0	14.9		565	8.3	7.5		
1973	1490	19.1	14.4		581	8.3	7.4	66.8	74.3
1974	1536	18.7	14.9		615	8.1	7.4		

[a] The birth rate is defined as the ratio of total births for a specific year to total midyear population.
[b] The total fertility rate is defined as average number of children born to a female in a given population (times 1000), based on a set of age-specific birth rates for a particular year.
[c] The death rate is defined as the ratio of total deaths for a specific year to total midyear population.
[d] Life expectancy is from age 0.

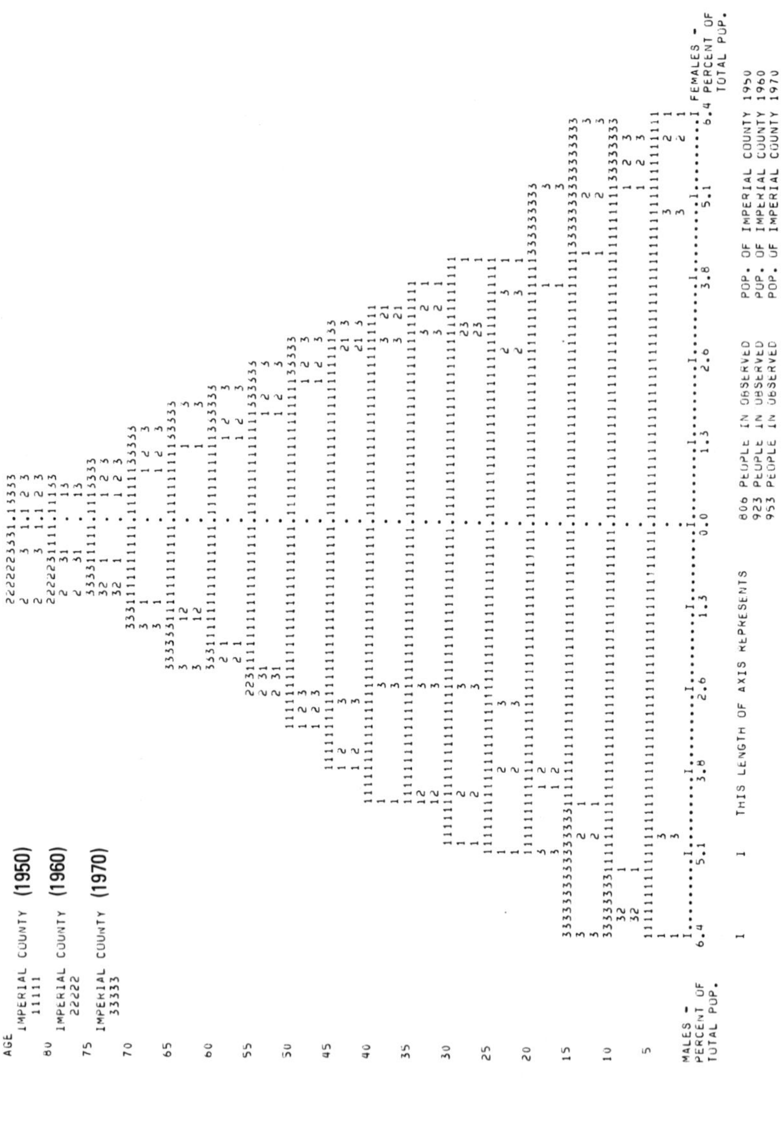

FIGURE 3-3. Comparative age structures for Imperial County, 1950–1970.

Population–Economic Data Analyses 95

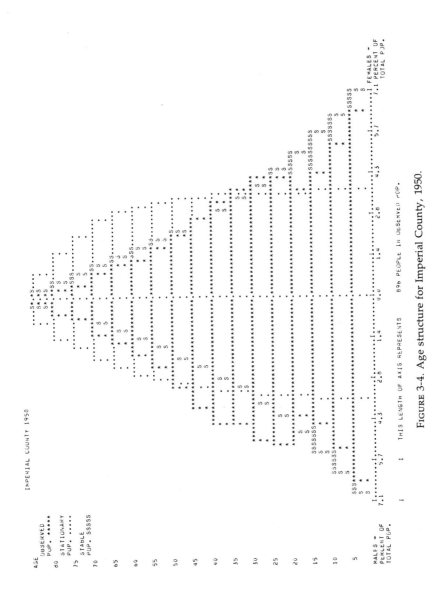

FIGURE 3-4. Age structure for Imperial County, 1950.

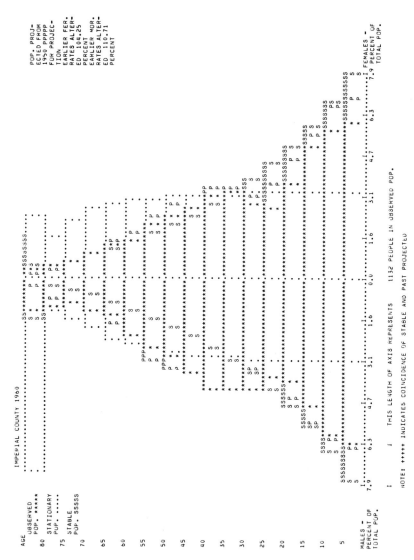

FIGURE 3-5. Age structure and age-specific migration for Imperial County, 1960.

Population–Economic Data Analyses

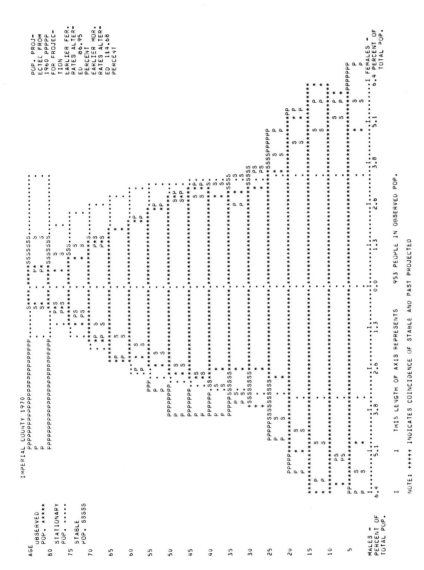

FIGURE 3-6. Age structure and age-specific migration for Imperial County, 1970.

muters, most of whom are male, the decrease from 1960 to 1970 appears to be a residential effect, with a maintenance of about a 1.2 sex ratio, if the population of the county were constituted of job holders and relatives of job holders (i.e., determined by place of work, rather than place of residence). Since the sex ratio for the non-Spanish–American population is 1.00 for 1970, unbalanced county sex ratios for 1950, 1960, and 1970 (adjusted) would appear to be due to the excess of Mexican agricultural labor. The implications for geothermal development of a 20% male excess are unclear, except that a larger labor pool is available than indicated by raw figures. If there is significant industrial development in the county, the traditional sex ratio will likely decrease, a residual of the agricultural character of a past labor force.

The 22% of persons aged 10–19 in 1970 undoubtedly caused many dependency problems, such as strains on educational institutions and families. However, this large "baby boom" age group (ages 19–28 in 1979) may be presumed to have partially out-migrated. The stationary (life table) population distributions in Figures 3–4 through 3–6 reveal better health conditions during the 1950s and 1960s. The stable 1970 pyramid implies a future tendency toward an increase of aged females, but a decrease in aged males.

Migration is revealed in Figures 3–5 and 3–6 by subtracting the observed from the past-projected population. To study age-specific migration between 1950 and 1970 it is convenient to aggregate ages by three larger categories: 0–29, 30–64, and 65+. Age structures for 1960 and 1970 are distorted by border commuters. Since these workers were nearly all males, comparisons of female age structures are more accurate. In 1960 female migration rates (based on past-projected population with direction indicated) were -14.3, -11.5, and $+10.0$ for aggregated ages. In 1970 these rates were -10.3, $+0.4$, and $+9.2$. Thus, female children and young women have been leaving the county, while retirement age women have been returning. The middle-aged segment has changed from strong out-migrating in 1960 to neutrality in 1970. Because of the youthful age pyramid, most migrants (counting in- and out-migrants equally) are in the 0–29 age category (65.2% in 1960 and 89.9% in 1970). The greatest female out-migration was for the ages of 20–24, with rates of -25.7 and -26.1 in 1960 and 1970. For the young, employment, educational, and life-style forces in more urban, sophisticated regions of the West appear to account for these significant departures. Because of possible data unreliability, males' rates were estimated for ages 0–19 and 65–79. In 1960, rates were -6.5 and $+12.0$ for them; in 1970, rates were -4.6 and $+9.7$. Hence, male migratory patterns also show the departure of the young and return of the old.

The age characteristics of net migration for the county are complicated further by in- and out-migration for specific segments of the population. The U.S. Census records only in-migration in the form of data on residence five years earlier. Table 3–5 shows that 10,691 persons in-migrated into the county in 1965–1970. Since there is a net out-migration rate of approximately 1% per year, persons in-migrating perhaps yield a gross out-migration rate of over 2%.

Table 3-5 shows that the Spanish-speaking population is much more stable geographically for movement within the U.S. (only 8.1% were in a different county in 1965 compared to 22.1% for anglos). This stability of movement in the U.S. is counterbalanced by a much higher international in-migration (8.8% versus 1.0% for anglos).

In summary, out-migration has been a key factor in the population dynamics of Imperial County. The implications for energy development of departure from the county of large numbers of young adults and associated children will be discussed later.

Family composition in Imperial County differs from that of Cali-

TABLE 3–5. IN-MIGRATION FOR SEGMENTS OF IMPERIAL COUNTY POPULATION, 1965–1970[a]

Residence five years earlier	Non-Spanish-speaking in-migration		Spanish-speaking in-migration	
	Rate (%)	Number	Rate (%)	Number
Same house	46.7	17,395	47.3	14,189
Different house	45.3	16,878	40.2	12,069
Same county	23.1	8,631	32.1	9,625
Different county	22.1	8,247	8.1	2,444
Same state	12.5	4,690	6.1	1,844
Different state	9.5	3,557	2.0	600
Region, if different state:				
North-East	0.7	268	0.0	5
North-Central	1.8	669	0.2	71
South	2.8	1,059	0.9	263
West	0.4	1,521	0.9	261
Abroad	0.1	360	0.9	2,629
Moved 1965 (residence not reported)	0.7	2,611	0.4	1,094

[a]Source: U.S. Bureau of the Census (1971).

fornia. In 1960 and 1970, the high fertility in the county is clearly manifest in the large percentages of families with young children under age 6 (in 1960, this percentage was 38.6 for Imperial County versus 30.9 for California; in 1970, these percentages were, respectively, 31.4 and 26.0). The somewhat lesser percentage of families with no children under 18 for Imperial than for the state implies greater family dependency in the county (in 1960 this percentage was 35.2 for Imperial County versus 42.6 for California; the respective percentages in 1970 were 37.5 and 44.3).

POPULATION ECONOMICS

Analysis of population composition and change reveals a region at variance with many demographic features of California, such as a balanced sex ratio, in-migration of younger persons, and population increase. The county economy, primarily based on agriculture, will be a critical element in geothermal development.

The income distribution of the county was examined from 1950 to 1970 by inflating all income categories by the ratio of the mean incomes for a ten-year period. To achieve a somewhat larger economic area for these inflators, Imperial's inflators were averaged with those for Los Angeles County. Incomes were inflated by 87% for 1950–1960 and by 53% for 1960–1970. Table 3–6 shows comparable income classes after multiplication by the above weights. The percent of top family incomes (above $28,611 in 1970) appears to be rather stable over 20 years. However, the second highest and lowest income brackets consistently decreased over the period, whereas the middle categories in-

TABLE 3–6. INCOME DISTRIBUTION FOR IMPERIAL COUNTY 1950–1970 WITH CLASSIFICATION SUBJECT TO INFLATION[a]

Income class 1950	Percent	Income class 1960	Percent	Income class 1970	Percent
10,000+	3.1	18,700+	3.7	28,611+	2.9
5,000–9,999	21.2	9,350–18,699	19.3	14,305–28,610	15.3
3,500–4,999	15.0	7,479–9,349	14.8	10,013–14,304	19.6
1,000–3,499	44.8	1,870–6,544	50.5	2,861–10,012	50.8
0–999	15.7	0–1,869	11.4	0–2,860	10.6

[a] Inflation of classifications calculated by average inflation of mean incomes for Imperial and Los Angeles Counties. Source: U.S. Bureau of the Census (1950–1970).

TABLE 3-7. PERCENT MALES IN EMPLOYMENT CATEGORIES FOR IMPERIAL COUNTY 1950–1970[a]

Employment categories	1950	1960	1970
Professional	4.1	4.8	8.7
Farmers and farm managers	9.2	6.3	4.8
Managers (nonfarm)	9.8	9.8	13.5
Clerical	3.0	2.9	4.7
Sales	4.4	4.0	5.7
Craftsmen, foremen	11.7	11.2	17.5
Operatives	14.3	12.3	14.4
Private household	0.0	0.0	0.0
Other service workers	4.7	4.2	8.0
Farm laborers and foremen	30.9	36.9	15.9
Other laborers	6.4	4.2	6.2
Not reported	1.0	2.8	0.6
Aggregated categories			
Farm (including farm managers)	40.1	43.2	20.7
Clerical and sales	7.4	6.9	10.4
Professionals and managers	13.9	14.6	22.2
Craftsmen and operatives	26.0	23.5	31.9
Other	12.6	11.8	14.8

[a] Source: U.S. Bureau of the Census (1950–1970).

creased by 10%. This trend toward middle-class categories also appears for male employment (Table 3-7); there is an increase for 1950–1970 of 11% in clerical and professional–managerial employment. (For California, this increase was only 4%.)

The industrial classifications for Imperial County, Kern County, and California appear in Table 3-8. These can best be analyzed by separating occupational divisions by sex, and by eliminating the agricultural component entirely. For males in Imperial County, the most significant change in industrial classification is an 8% rise in the public administration and government education (versus a 2.8% increase in these for California and a 3.7% increase for Kern). Again, this increase is reflective of growth in the middle class. For males, a significant difference is a county manufacturing component only 57% smaller than the state, but a greatly increased utility and sanitary category (304% larger than the state). For females in Imperial, the most important longitudinal trend is an increase of 4% in government education workers. On the female side, compared to California, there is an even more sharply reduced manufacturing component (77% less

TABLE 3–8. INDUSTRIAL DISTRIBUTIONS (IN PERCENT) FOR IMPERIAL COUNTY, KERN COUNTY, AND CALIFORNIA, 1950–1970[a]

	1950		1960		1970	
Variables	Male	Female	Male	Female	Male	Female
Imperial County						
Number of workers	18,599	4,518	21,613	6,414	15,397	8,082
Labor sex ratio	4.12		3.37		1.90	
Industries						
Agriculture	43.4	8.6	48.4	6.3	25.1	8.8
Construction–Mining	6.6	0.6	4.3	0.4	7.5	1.0
Manufacturing	7.1	2.0	6.6	1.9	8.5	3.2
Utilities–Sanitary	4.8	2.0	3.6 (A)	4.1 (A)	5.5	3.5
Wholesale–Retail	17.8	33.0	14.7	28.3	22.3	28.7
Public administration	3.9	4.9	4.5	5.5	8.0	5.5
Education (government)	1.0	8.7	1.7	10.7	4.2	12.6
Other	15.6	40.8	14.9	47.3	16.6	36.7
Kern County						
Number of workers	58,065	18,287	69,262	29,011	71,991	37,548
Labor sex ratio	3.17		2.39		1.92	
Industries						
Agriculture	22.0	8.1	19.1	5.5	15.4	6.0
Construction–Mining	20.8	1.8	16.2	1.5	18.5	2.3
Manufacturing	9.5	3.0	12.2	3.9	9.8	3.3
Utilities–Sanitary	1.9	0.6	1.6 (A)	1.3 (A)	2.0	1.3
Wholesale–Retail	16.4	28.8	15.8	25.3	19.4	25.9
Public administration	8.4	8.7	9.7	7.9	10.4	6.8
Education (government)	1.8	11.3	3.1	12.0	3.8	13.7
Other	30.5	45.0	29.2	44.7	20.7	40.7
California						
Labor sex ratio	2.40		2.03		1.64	
Industries						
Agriculture	9.9	2.1	6.2	1.3	4.2	1.2
Construction–Mining	11.6	1.0	9.5	1.1	8.7	1.1
Manufacturing	21.5	15.0	27.5	17.2	25.4	15.1
Utilities–Sanitary	2.1	0.7	1.5 (A)	1.0 (A)	2.2	0.7
Wholesale–Retail	20.8	26.0	17.6	21.0	20.6	21.7
Public administration	6.6	6.6	6.5	5.3	7.2	5.1
Education (government)	1.6	5.7	2.7	7.7	5.0	9.6
Other	25.9	42.9	28.5	45.4	27.7	45.5

[a] Note: (A) indicates sex distribution estimated from 1970 data. Source: U.S. Bureau of the Census (1950–1970).

than the state). Again, the utility work force is larger. In summary, eliminating the effects of agriculture and sex ratio, two agricultural counties, Kern and Imperial, appear highly similar in industrial distribution but contrast with the state.

The above analysis controls for the influence of the labor force sex ratio. For the state, this ratio fell from 2.40 in 1950 to 2.03 in 1960 and finally to 1.64 in 1970. For Imperial County, the figures were 4.12, 3.37, and 1.90, respectively. It is assumed that the labor force sex ratio will continue to decrease. Thus, in considering labor market potential for geothermal-related industrialization, an important source of labor, presently only partially tapped, is females. The large gap with the state in female manufacturing employment appears particularly important.

Imperial County has higher rates for poverty and unemployment than Kern and the state. Unemployment rates in 1970 for Imperial, Kern, and California were respectively 7.1%, 4.2%, and 4.2% for males age 14 and older and 6.4%, 2.9%, and 2.8% for females 14 and older. These figures likely reflect agricultural unemployment. In 1970 16.1% of Imperial County families were below poverty level versus 12.6% for Kern and 11.9% for California. Increased poverty may reflect the large number of agricultural laborers.

Housing and Transportation

Space and crowding factors in Imperial County housing are given in Table 3–9. A high vacancy rate is apparent, but may be due to seasonal factors. In room availability, the county slightly lags statewide trends. Thus, it is not surprising that county residents are more crowded than the state, although Imperial appears to have been catching up relatively since 1950.

For value of housing, an analysis was performed which inflated home categories over the two decades by the consumer price index. Table 3–10 reveals a consistent lag of the county behind statewide home prices. The sharp increase in the highest-priced homes in Imperial from 1950 to 1960 is unexplained. Although behind the California increase in value in the 1950s, for the 1960s, Imperial's homes experienced an increase in value of 124% compared to 95% for the state.

Data on air conditioning, heating equipment, and appliances are shown in Table 3–11. As expected for the climate, the county has three times more air conditioning usage than the state. Decreased heating needs are shown by comparisons of the aggregated category of room

TABLE 3-9. SIZE AND CROWDING IN HOUSING (IN PERCENT) FOR IMPERIAL COUNTY AND CALIFORNIA, 1950-1970[a]

Variables	1950		1960		1970	
	Imperial	California	Imperial	California	Imperial	California
Population/occupied unit	3.6	3.0	3.6	3.1	3.5	3.1
Housing units	17,904	—	21,916	—	23,401	—
Owner occupied	43.8	50.4	46.8	53.2	51.9	51.6
Renter occupied	48.4	42.3	37.4	37.9	37.8	42.3
Vacant year round	5.9	4.9	12.4	8.8	9.2	5.7
Number of rooms						
1-2	24.6	14.6	18.8	10.9	10.4	8.6
3-4	40.5	38.8	37.8	35.8	39.2	36.4
5-6	24.8	28.7	36.1	44.9	40.1	42.3
7+	5.1	7.7	7.1	8.2	9.2	12.5
Median number rooms						
Crowding in occupied units Persons/Room:						
1.00	65.6	89.5	78.3	92.2	80.6	92.0
1.05-1.50	13.4	6.0	21.6	6.2	11.8	5.5
1.51	17.3	3.0	21.6	1.4	7.5	2.3
Median number persons	3.2	—	3.2	—	3.0	—

[a] Source: U.S. Bureau of the Census—Census of Housing (1970).

TABLE 3–10. HOUSING VALUES (IN PERCENT) FOR IMPERIAL COUNTY AND CALIFORNIA, 1950–1970, WITH CLASSES SCALED FOR INFLATION BY CONSUMER PRICE INDEX[a]

	1950			1960			1970	
Housing classes	Imperial	California	Housing classes	Imperial	California	Housing classes	Imperial	California
0–5,999	57.6	17.2	0–7,499	34.3	11.8	0–9,999	28.1	4.7
6,000–9,999	23.5	36.5	7,500–12,499	33.5	38.1	10,000–14,999	28.4	10.7
10,000–19,999	16.5	39.2	12,500–24,999	24.4	26.0	15,000–34,999	38.7	67.0
20,000+	2.2	7.0	25,000+	7.5	3.8	35,000+	4.6	17.3
Median value	5,245	9,564	—	6,200	12,500	—	13,900	24,400

[a]Source: U.S. Census of Housing (1970).

TABLE 3-11. AIR CONDITIONING AND HEATING EQUIPMENT (IN PERCENT) FOR IMPERIAL AND KERN COUNTIES AND THE STATE, 1970[a]

Variables	Imperial	Kern	California
Percent of units air conditioned	77.2	34.6	24.9
Heating equipment			
Steam or hot water	0.2	1.7	3.1
Warm air furnace	25.2	30.7	35.1
Built-in electric units	9.0	2.7	6.7
Floor, wall, or pipeless furnace	11.0	42.9	32.1
Room heaters	29.4	17.2	18.3
Fireplaces, stoves, or portable heaters	19.8	3.7	3.6
None	4.9	0.8	0.8
Appliance			
Washing machine	65.6	71.6	65.5
Clothes dryer	16.9	40.0	41.8
Dishwaster	12.1	22.7	26.8
Food freezer	21.3	26.7	20.9
TV	81.3	87.8	94.7

[a] Source: U.S. Census of Housing (1970).

heaters, fireplaces, stoves, portable heaters, and no heaters. This combined category was 54.1% for county housing in 1970, versus 22.7% for the state. Also, Imperial has rather reduced appliance utilization. The present tradeoff between increased air conditioning and decreased home heating and appliance usage will likely be considerably altered, if multipurpose use of geothermal energy were fully exploited. Geothermal energy has been used elsewhere for space heating as well as air conditioning.

The major housing fuel categories for the county were utility gas (used in 54.4% of households); electricity (31.9%); bottled, tank, or LP gas (7.3%); no fuel (4.5%); and other (1.9%). The major differences from the statewide pattern are reduced use of utility gas and increased use of electricity. Possible explanations are the low price of electricity due to the relatively large component of hydroelectricity in the county and the limited availability of gas connections.

Transportation statistics for Imperial were compiled only in 1960. Major categories for transport to work were private auto or car pool (69.9%), bus or streetcar (7.1%), walking (8.0%), and working at home (6.2%). These are extremely close to the statistics for the state and reveal a predominance of automobile transportation. Auto transportation is estimated to be currently in the range of 80–85%.

Regional Socioeconomic Comparisons

In order to describe important county trends, socioeconomic characteristics relevant to energy were studied by computer mapping. Data for county enumeration districts (henceforth called EDs) were obtained from the fifth count of the 1970 Census (U.S. Bureau of Census, 1974). These data were displayed using Automap II, a standard computer mapping program (Environmental Systems Research Institute). Five levels of data values were chosen, with extremes (level one and level five) approximately eighths or sixteenths. Since 58.6% of Imperial's population resides in the towns of Brawley, Calexico, and El Centro, and another 35.7% lives in the central irrigated valley, the central valley and the largest three towns were mapped separately. The 5.7% of the population residing outside the central valley on the East and West Mesas is not represented on the maps.

The location of geothermal energy may be approximated by areas called Known Geothermal Resource Areas (KGRAs). Figure 3–7, a map of percent Spanish speaking for the central valley, has superimposed on it a map of the three central valley KGRAs. For all analyses, an ED was identified as geothermal if more than 50% of its surface area was contained within a KGRA. In addition, because of their size and location just outside the Heber KGRA border, EDs in the towns of Calexico and El Centro were identified as geothermal. Although not mapped, an ED in the lower East Mesa contains a promising KGRA which is being actively drilled by Republic Geothermal Corporation. However, this ED contains extensive land area and few of its 259 residents are likely to be affected; hence, it was not identified as geothermal.

Using computer mapping technology, regional and geothermal results are presented for the following characteristics: Spanish language, dependency ratio, labor force availability, mobility, income, and rental market over $100. Spanish language residents show a concentration in Calexico and periphery (31% of SAs in the county) and in Brawley and periphery (22.9%). Thus, in the Heber KGRA and Brawley–Salton Sea KGRAs, the Spanish-speaking are of particular importance. Within towns, a directional axis of SA composition is apparent: southeast (high) to northwest (low) for Brawley; northeast (high) to southwest (low) for El Centro; and south (high) to north (low) for Calexico. In drilling and geothermal plant locations near these towns, such well-defined axes may be a consideration.

A dependency ratio (persons aged under 18 plus persons 65 and over/persons 18 to 64) was calculated and mapped as shown in Figures 3–8a, 3–8b, 3–8c, and 3–8d. This measure is ethnically related because

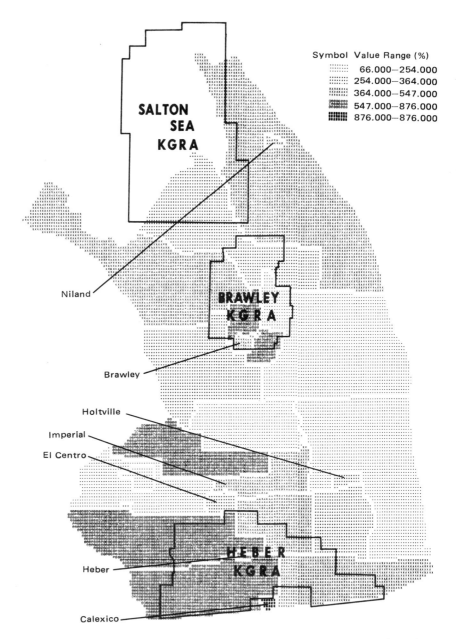

FIGURE 3-7. Percent Spanish-speaking, County of Imperial (ED and KGRA boundaries superimposed).

Population–Economic Data Analyses 109

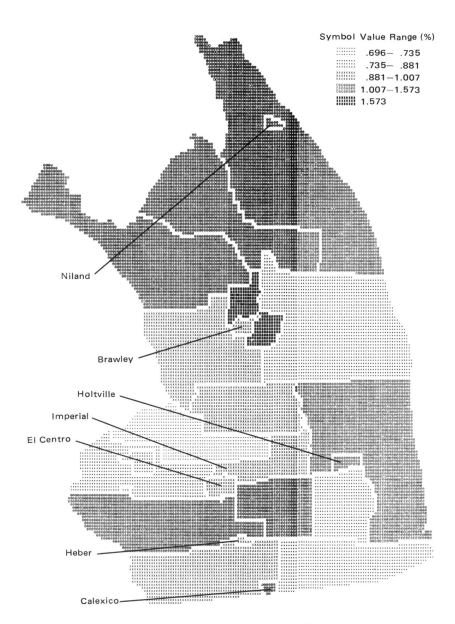

FIGURE 3-8a. Dependency ratio, County of Imperial.

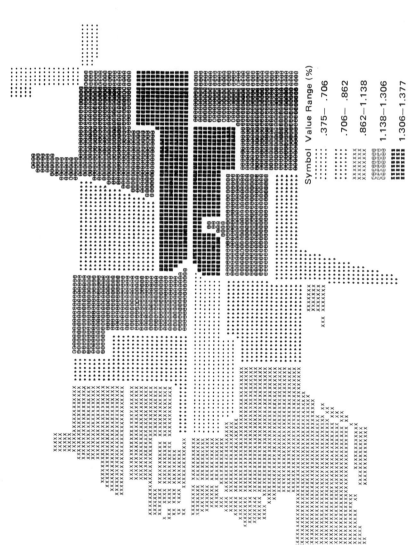

FIGURE 3-8b. Dependency ratio, City of Brawley.

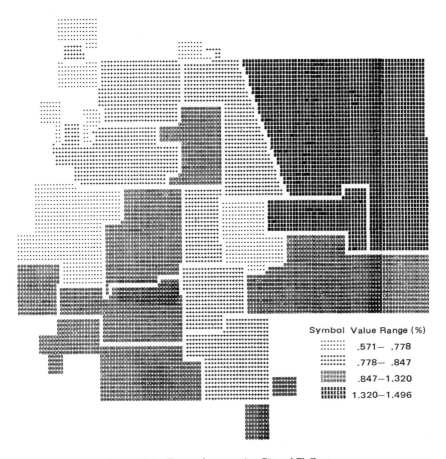

FIGURE 3-8c. Dependency ratio, City of El Centro.

of the very large portion of children and adolescents in the SA population. Thus, dependency closely corresponds to percent SA ($r = 0.44$), with a greater than average concentration in KGRAs. (The correlation coefficient, abbreviated as r, measures positive association between two variables on a scale of 0 (no association) to 1 (the variables vary together), or negative association between two variables on a scale of 0 (no association) to -1 (the variables vary in an opposite manner). Directional axes for the towns are not apparent for dependency, but areas of highest concentration are present in east-central Brawley, northeast El Centro, and north Calexico. The opposite locations of high dependency and high percent MA for Calexico are due to the

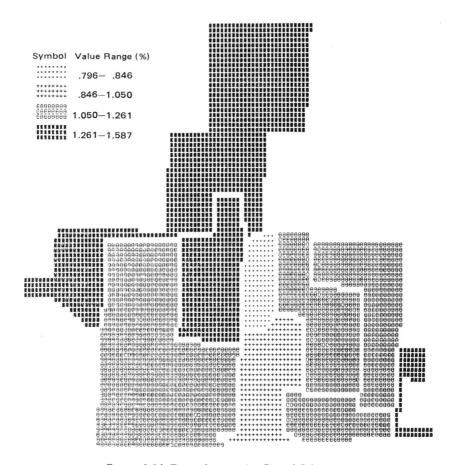

FIGURE 3-8d. Dependency ratio, City of Calexico.

large percentage of young family housing for both ethnic populations in north Calexico.

Male labor force availability shown in Figures 3–9a, 3–9b, 3–9c, and 3-9d is defined as the number of males 17–64/total number of males. By definition, it is inversely related to dependency, although the latter includes both sexes. For the northern combined KGRA, male labor force availability is very low in peripheral areas, but high in Brawley. For the southern KGRA, a medium level is apparent with El Centro medium and Calexico low. Brawley, which contains the preponderance of persons in the northern geothermal areas, clearly offers, in addition, an age structure favoring male work availability. Within

Population–Economic Data Analyses

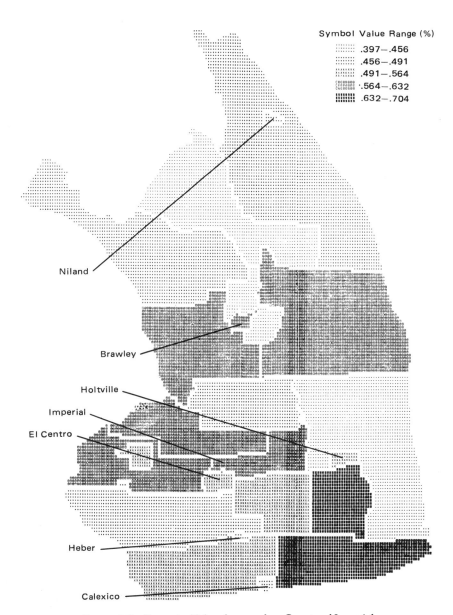

FIGURE 3-9a. Percent of labor force males, County of Imperial.

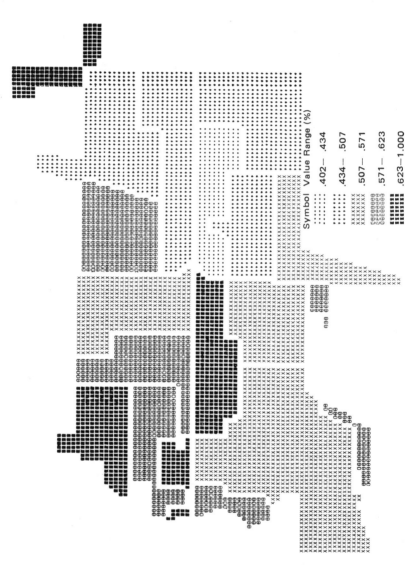

FIGURE 3-9b. Percent of labor force males, City of Brawley.

Population–Economic Data Analyses

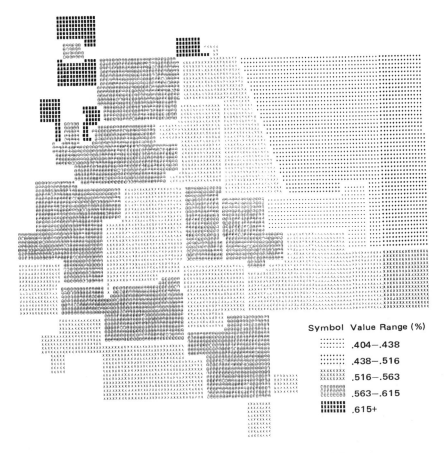

FIGURE 3-9c. Percent of labor force males, City of El Centro.

the towns, this characteristic is highest in the west-central part and two northern corners of Brawley, the northeast corner of El Centro, and the southwest corner of Calexico.

As shown in Figure 3–10, female labor force availability is very highly correlated ($r = 0.58$) with that of males. Major differences are the reduction in female relative to male level in the highest income areas of southwest Brawley, El Centro, and north Calexico. A similar reduction in the southwest part of Calexico, an area of farm labor activity, may be due to a reduced availability of housing for married couples.

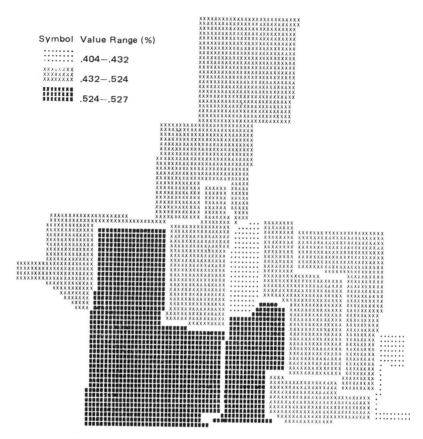

FIGURE 3-9d. Percent of labor force males, City of Calexico.

MIGRATION AND INCOME CHARACTERISTICS

In-migration to the county was studied by EDs for 1965–1970. For four towns, the rates of in-migration into the county for 1965–1970 were 7.3% for Brawley, 10.8% for El Centro, 18.6% for Calexico, and 11.8% for Niland. El Centro leads in interstate migration as one would expect for the major county town. Calexico is characterized by the highest residential, intercounty, and foreign in-migration, but the second lowest interstate in-movement. In the central business district (ED 68), the in-migration rate from abroad equals 29.7%. These figures for Calexico are even more remarkable when one recalls that the county, including this town, has been subject to about a 1% out-migration per

Population–Economic Data Analyses 117

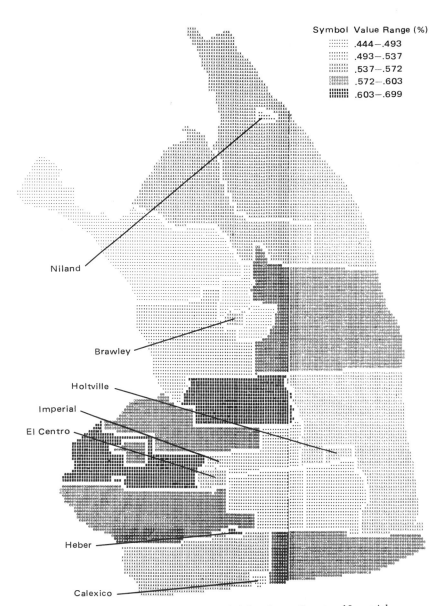

FIGURE 3-10. Percent of total female labor force, County of Imperial.

year. The town has balanced its population at a level of about 10,000 by flows in and out.

Regional mobility patterns were studied by 1965 residence in a different house (household mobility), different county (county mobility), and different state (state mobility). These data illustrated in Figures 3–11, 3–12, and 3–13, are for in-movement rather than out-movement. Hence, for a county in which out-movement has predominated, the bulk of total moves is ignored. However, since town populations were very stable between 1960 and 1970, it may be assumed that regional out-movement levels are highly correlated with in-movement levels. These data show that mobility up until now does not correspond to KGRA boundaries. Household mobility was in rural and peripheral areas in the eastern third and southern third of the central valley, except for the southeast corner; relatively high in the northeast third; and medium to low in major towns. County mobility in nonurban areas in the northwest third is low, with moderate levels for the remainder; the cities of Calexico and Brawley had moderate rates while El Centro was high.

Since interregional mobility has been found to be an important correlate of other social variables, including vital rates and prospective mobility (Butler, 1969; Pick, 1974), planners concerned with the labor force in El Centro should note the high mobility, with a decreasing gradient from west to east. Since interstate in-migrants compose only 30% of intercounty inmigrants, they are of less importance. The most notable feature, again, is elevated levels for El Centro and adjacent areas to the east and west.

Regional income differentials are very sharply defined, with higher incomes in El Centro, western rural El Centro, Holtville, and rural Calexico to the east. Rural parts of the Salton Sea and Brawley KGRAs and the town of Heber are at the lowest level of income. Some people in the Salton Sea area are transients and/or intermittent visitors, Figure 3–14. A general countywide gradient for rural areas runs from north (high) to south (low). Clear income axes exist in Brawley (west–high to east–low), and in Calexico (north–high to south–low). Regionally, the location of rentals in the rental market above $100 corresponds very closely to income: $'r$ (percent rentals from $100 to $149 in the rental market, percent family income from $9,000 to $25,000) = -0.48, while r (percent rentals from $200 to $249, percent family income from $9,000 to $25,000) = 0.04.

In summary, the county appears to have well-defined regional patterns. In geothermally active regions, such information may contribute to future industrialization patterns related to energy development.

Population–Economic Data Analyses

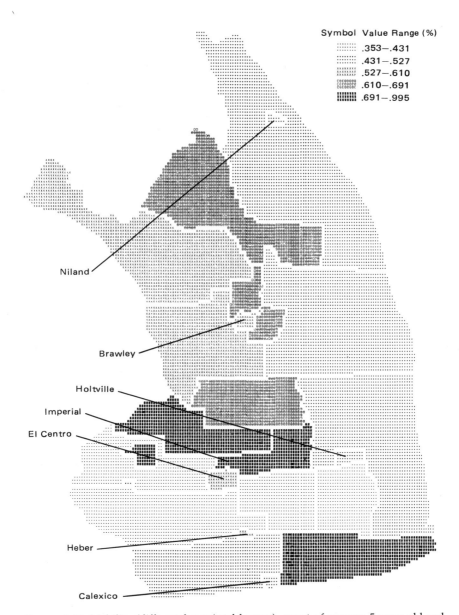

FIGURE 3-11. Mobility (different house/total houses), count of persons 5 years old and over by residence in 1965.

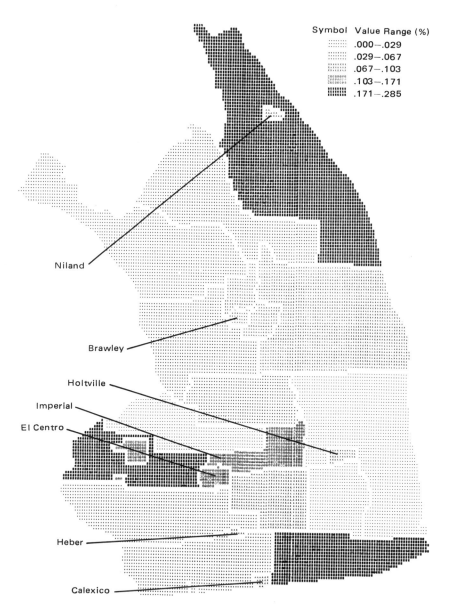

FIGURE 3-12. County mobility (county migration/total houses), count of persons 5 years old and over by residence in 1965.

Population–Economic Data Analyses

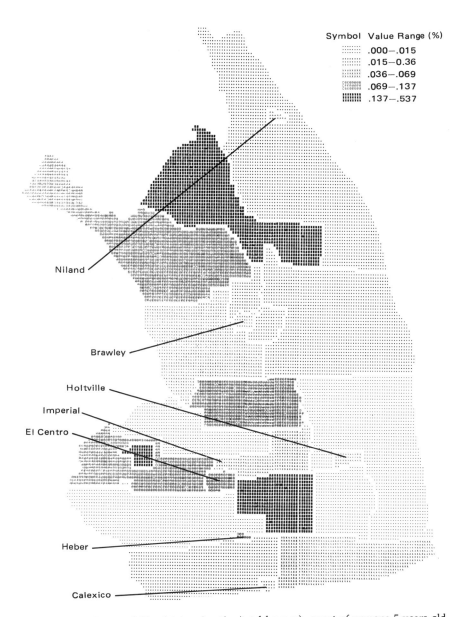

FIGURE 3-13. State mobility (state migration/total houses), count of persons 5 years old and over by residence in 1965.

FIGURE 3-14. Trailer camp, southern edge of Salton Sea, Imperial County, 1978.

DISCRIMINANT ANALYSES OF GEOTHERMAL AREAS

Previous sections have analyzed aggregated county population–economic data and regional county data. Discriminant analysis was performed in order to distinguish geothermal and nongeothermal regions of Imperial from each other. Discriminant analysis answers the question, "Of all variables under consideration, which one (or which several together) is (are) the best to use in distinguishing the regions from each other?" The potential usefulness of such results in geothermal development is that characteristics distinguishing recipients of localized effects of energy development, such as noise, smell, and, to a certain extent, nonelectrical uses can be more accurately estimated.

In discriminant analysis variables (say three in number) are chosen to do the discrimination (these are $x_{1_*}, x_{2_*}, x_{3_*}$, where the asterisk stands for either A (for sample unit A) or B (for sample unit B) (Lachenbruch, 1975; Eisenbeis and Avery, 1972). The sample units in the present study are EDs. For sample units A and B, the following equations may then be established:

$$y_A c_1 x_{1A} + c_2 x_{2A} + c_3 x_{3A} \qquad (1)$$
$$y_B c_1 x_{1B} + c_2 x_{2B} + c_3 x_{3B} \qquad (2)$$

The c's are constants which are chosen to obtain y_A as different in value from y_B as possible. The x's are referred to as independent variables.

Population–Economic Data Analyses

In stepwise discriminant analysis, which is used here, one begins with a long list of candidates for independent variables, say $x_1, x_2, x_3, x_4, x_5, x_6, x_7,$ and x_8 (examples of x's from the Imperial study are percent below poverty level and percent now married). First, one x is chosen (with corresponding c) for equations (1) and (2); then two are chosen; then three, and so on until all eight are chosen. One then picks out (to put it simply) equations having the least number of x's which give good results on distinguishing the y's.

After one has chosen discriminant equations, one may use these with the data at hand to see how good a job they do in classifying sample units (e.g., for EDs, classification as a geothermal or nongeothermal region). The apparent error rate gives the fraction of sample units (EDs) in a region (e.g., geothermal, nongeothermal) which were erroneously classified.

Geographical identification of EDs was accomplished by superimposing KGRA over ED boundaries. The East Mesa KGRA was omitted from this analysis because it is contained within a single ED (ED 2) which is spread over such a large land area that less than 50% of the land area is within the KGRA.

Three problems complicate the above identification: (1) fractional boundary overlaps, (2) lack of sufficient number of EDs in a KGRA, and (3) urban population irregularities. Boundary overlaps were resolved as follows: If more than 50% of an ED is contained within a KGRA, it is classified as being geothermal and belonging to that KGRA. Figure 3–7 shows that the Salton Sea KGRA contains only one ED, according to the above criterion. There are no definite rules in discriminant analysis on the minimum number of sample units in a classification category, but a total of one, as in the Salton Sea case, is considered too small. This problem was resolved by combining KGRAs in cases where the number of sample units in a KGRA is considered too small. Thus, the Salton Sea sample units are sometimes combined with the Brawley sample units, and the resultant larger unit henceforth is referred to as the Brawley–Salton Sea CKGRA (consolidated KGRA). In some analyses, all three KGRAs are combined together, and the region containing EDs for all three KGRAs is henceforth referred to as geothermal. Non-KGRA EDs, outside urban areas, are referred to as nongeothermal.

Urban population irregularities are due to the circumstance of location of the major towns (Brawley, El Centro, and Calexico) immediately inside or immediately outside the boundary lines. The U.S. Geophysical Survey included Brawley and El Centro within KGRAs, but excluded Calexico by a narrow margin due to a slight irregularity in

TABLE 3–12. GEOGRAPHICAL DISTRIBUTION OF SA POPULATION[a]

Locality	Population	SA population	Percent SA	Percent of population in county	Percent of SA population in county
Urban:					
Holtville	3,628	992	27.3	4.8	2.8
Brawley	13,746	6,549	47.6	18.4	19.1
El Centro	19,272	6,650	34.5	25.8	19.4
Calexico	10,625	9,701	91.3	14.2	28.3
	47,271	22,900	48.4	63.4	69.6
Heber	710	578	81.4	0.9	1.6
Rural:					
Holtville	1,819	441	24.2	2.4	1.2
Brawley	3,069	1,317	42.9	4.1	3.8
El Centro	6,748	2,443	36.2	9.0	7.1
Calexico	1,453	928	63.8	1.0	2.7
	13,089	5,129	39.1	17.4	14.8
Other population	13,422	4,661	34.7	18.0	13.6

[a] Source: U.S. Bureau of the Census (1971).

the boundary. As shown in Table 3–12, such a large proportion of county population is contained within each of these three towns that it was considered uniform treatment either to exclude all three towns from KGRAs or to include all three in KGRAs. In the following analyses, both uniform classifications were used. For samples referred to as entire county, all three towns were included within KGRAs. For samples referred to as rural, the three towns were excluded from the analysis.†

Independent variables included as eligible for the stepwise discriminant analysis are shown in Table 3–13. Except for variables 19–21 and 25–27, these were chosen because they were important in the

†For the entire County sample, the regions are composed of the following EDs: (1) Salton Sea CKGRA (EDs 13–28, 30, 8); (2) Heber KGRA [EDs 65–73 (including 65B), 74, 75, 43–57 (including 55B), 58, 63]; and (3) nongeothermal (EDs 29, 31, 4–7, 1–3, 59–62, 64, 32–35, 36, 37, 38–40, 41, 42, 9–12, 76–78). For the rural sample, the regions are composed as follows: (1) geothermal (EDs 30, 8, 74, 75, 58, 63); and (2) nongeothermal (EDs 29, 31, 4–7, 1–3, 59–62, 64, 36, 37, 41, 42, 9–12, 76–78). The rural sample thus excludes all EDs within the city limits of Brawley, El Centro, Calexico, Holtville, and Imperial.

Population–Economic Data Analyses

TABLE 3–13. LIST OF VARIABLES IN DISCRIMINANT ANALYSIS

1. Percent Spanish speaking
2. Dependency ratio [persons (−18) + persons (65+)/persons (18–65)]
3. White fertility ratio
4. Spanish fertility ratio
5. Percent labor force age males (17–64) of all males
6. Percent labor force age females (17–64) of all females
7. Percent of population now married
8. Percent of persons (5+) in a different house in 1965
9. Percent of persons (5+) in a different county in 1965
10. Percent of persons (5+) in a different state in 1965
11. Percent of persons (5+) abroad in 1965
12. Percent unemployed (white) of white population 18 and over
13. Percent unemployed (Spanish) of Spanish population 18 and over
14. Percent of labor force, white collar
15. Percent of labor force, blue collar
16. Percent of labor force, farm laborers and foremen
17. Percent of families with income less than $4,000
18. Percent of families with income $10,000–24,999
19. Percent of housing units with 3 or less persons per unit
20. Percent of housing units with 1 or fewer persons/room
21. Percent of housing units (Spanish) with 1 or fewer persons/room
22. Percent of housing units with utility gas from underground pipes as heating fuel
23. Percent of housing units with bottled tank or LP gas as heating fuel
24. Percent of housing units with electricity as heating fuel
25. Percent renter-occupied units in the $100–250 rental market with rents $100–149
26. Percent renter-occupied units in the $100–250 rental market with rents $150–199
27. Percent renter-occupied units in the $100–250 rental market with rents $200–249

analysis of crude county data. Density variables (19–21) were chosen because density has proven an important population variable in recent studies of vital rates (Pick, 1976), crime (Galle et al., 1973), and other variables. Rental variables (25–27) were included because income was important in previously discussed crude county data—rental data provide additional refinement on household economic status.

The source of the data was the Fifth Count of the U.S. Census (U.S. Bureau of the Census, 1974)—the same data source used in the mapping. After sorting and aggregation by several computer routines, the data were analyzed by a standard stepwise discriminant analysis program—Biomed BMDP-7M (Dixon, 1976).

Results of the analysis are given in Table 3–14. For the entire sample with two regions (geothermal and nongeothermal), the distinguishing variables are utility gas, rents $200–$249, and percent Spanish speaking. The numbers in parentheses in this table are known in

TABLE 3-14. DISCRIMINANT ANALYSIS OF GEOTHERMAL REGIONS BY SOCIOECONOMIC CHARACTERISTICS

Grouping	Variables included (F to remove in parentheses)	F matrix		Apparent error rate	
Entire county [80][a] Two groups	Utility gas (27.8[b]) Rents $200–249 (11.5[c]) Spanish-speaking (11.1[c])		Geothermal [48]	Geothermal	0.167
		Nongeothermal [32]		Nongeothermal	0.281
Entire county [80] Three groups	Utility gas (15.4[b]) Spanish-speaking (3.6[d])	Heber [30] Nongeothermal [32] (Approximately $F = 11.2$)	Salton [18] Heber [30]	Salton Heber Nongeothermal	0.389 0.467 0.344
Rural portion [31] Two groups	Spanish-speaking (12.9[b])	Nongeothermal [25]	Geothermal [6]	Geothermal Nongeothermal	0.333 0.120
Rural portion [31] Two groups[f]	Income $9,000–24,999 (8.5[c]) Bottled gas (5.6[e])	Nongeothermal [25]	Geothermal [6]	Geothermal Nongeothermal	0.333 0.240

[a] [ ] = number of ED's.
[b] Significant at 0.9995
[c] Significant at 0.99
[d] Significant at .95
[e] Significant at .975
[f] Spanish-speaking variable excluded

statistics as F values; subject to certain assumptions, these give an indication by their size of the relative importance of the independent variables in distinguishing regions (Lachenbruch, 1975).

For the entire sample, the above three variables may be important, because they are good indicators of urban–rural differences. For the most important of these—utility gas—the urban–rural difference is very marked (especially in the towns and adjacent areas of Brawley and El Centro and in the town of Heber). Such a difference is clearly the result of a tendency for installation of gas transmission lines in more concentrated urban areas, in contrast to greater use of electricity in rural areas. Rentals of $200–$249 (as a percent of the $100–$249 rental market) also show greater concentration in urban areas than in rural (although the high values for this variable are in two eastern rural EDs). Percent Spanish speaking is very high in the Heber KGRA.

The importance of utility gas, higher rentals, and ethnicity is possibly due to a combination of geologic and human settlement reasons. Geothermal fields are contained in the Salton Trough, which extends from the Gulf of California to north of the Salton Sea, and are more likely to be on the central part of the trough.

Likewise, the irrigated valley is centrally located in the trough; and from a transportation and labor force standpoint, the county's towns were historically more likely to be centrally settled in the valley. Hence, the significance of the utility gas and rental variables stems from the concentration of utility gas and higher rents in urban areas. SA preference for urban residence leads to the importance of the SA variable (Grebler *et al.*, 1970).

When the entire county is analyzed for three regions (Brawley–Salton Sea CKGRA, Heber KGRA, and nongeothermal), again the urban–rural variables of utility gas and percent Spanish speaking distinguish the regions. These findings are somewhat confused by the clear discrimination between nongeothermal and Salton, and between nongeothermal and Heber, but lack of significant distinction between Salton and Heber. The latter lack of differentiation cannot be attributed to the central versus noncentral thesis, and likely stems from lack of significant north–south socioeconomic gradients.

For the rural analysis, the key variable distinguishing geothermal and nongeothermal regions is percent Spanish speaking. The explanation is twofold: First, nonurban EDs in KGRAs are weighted toward the Heber KGRA, a border region with a higher percent Spanish speaking. Second, the northeast adjacent ED to Brawley (ED 30) has a very high percent Spanish-speaking, so that one may discern on the map a central–noncentral gradient, less pronounced than that for utility gas in the entire county. Thus, the same geological–settlement explanation which was given earlier may be applied to the rural sample.

When the Spanish-speaking variable is excluded from the analysis for the rural sample, percent of persons with income from $9,000 to $24,999 takes on the greatest importance. The map of this income pattern reveals an inverse effect between the geothermal nonurban EDs (low value for this variable) and the nongeothermal EDs (high values). The explanation of this income gradient is not clear but is assumed to be an unusual characteristic of one particular county—Imperial. The variable bottled gas appears to be another nonurban variable—low in urban areas, high in rural—which is distributed on the central–noncentral gradient, even with exclusion of urban EDs.

Applications to Geothermal Energy Development

The discriminant analysis revealed significant socioeconomic difference between nongeothermal and geothermal regions, whether urban population is included or excluded. Some geothermal effects (such as noise) will be localized on KGRA land; others will be partly or fully dispersed from KGRAs (an example of the latter is electrical consumption in other counties resulting from grid transmission). There are effects, such as nonelectrical applications, which may or may not be located on KGRA land since up to 50 km pipeline transmission of the hot water is possible, such as in Iceland.

It is for localized effects that the results of the discriminant analysis are of potential use. For example, planners or agencies may be interested in the socioeconomic characteristics of noise recipients in comparison to nonrecipients. For a thorough agency or county plan, such results refine the data on recipients beyond either crude countywide or individual town data. At a future date when power plant sitings become clear, planners might wish to refine the ED data further, based on exact sites.

For the entire county sample, the indicator variables chosen distinguish much better between geothermal and nongeothermal than between the Salton CKGRA and the Heber KGRA. Thus, for localized effects, it may be wiser to construct policy based on socioeconomic differences solely between geothermal and nongeothermal regions, and to ignore individual KGRAs.

For nonlocalized effects (such as county per-capita electrical supply) planners, regulatory agency officials, and others are better off using countywide data. Intermediate effects (for example, nonelectrical employment) cannot yet be pinpointed regionally, so they present the greatest policy and planning difficulty. For the present, one suggestion

Population–Economic Data Analyses

TABLE 3–15. PERCENT OF POPULATION OF IMPERIAL COUNTY AND THE STATE FOR VARIOUS DESIGNATIONS IN 1970[a]

Designations	Imperial County	California
Mexican foreign stock	30.5	3.5
Spanish mother tongue	41.3	10.7
Spanish language	44.9	13.7
Spanish language or surname	46.0	15.5

[a] U.S. Bureau of the Census (1971).

might be to divide such intermediate effects by some percentage formula into two regions, geothermal and nongeothermal. Then, policy could be based partly on localized and partly on countywide data results.

SPANISH-AMERICAN POPULATION OF IMPERIAL COUNTY

The 1970 population of Imperial was composed of 46% SAs, using the U.S. Census (1971) definition of Spanish language or Spanish surname. In Table 3–15, Imperial County is compared to California for these various designations.

Whichever definition is used, the county has a very large and significant SA component. It is important to distinguish the legal Mexican residents of Imperial County on the basis of citizenship. Those not citizens of the U.S., presumably citizens of Mexico, are referred to as legal Mexican aliens (abbreviated as LMA). These persons are counted in the U.S. Census if their place of residence is in the County.

The LMA population residing in Imperial is estimated from data in an unpublished report of the Immigration and Naturalization Service to be approximately 6,000 persons. Another group of LMAs are the border commuters who varied seasonally in number in 1970 between roughly 6,000 and 12,000 (estimated from U.S. Senate, 1971). Finally, the SA population includes illegal Mexican aliens (referred to as IMA) who have crossed the border illegally. There is no accurate estimate of IMA population in the county.

Resident SAs

Although in other regions of California there was an early history of Spanish and Mexican landholding, the creation of an irrigated valley only in 1901 precluded such 19th Century landholding. Migration from Mexico to the United States increased from then to a peak in the 1920s when about 500,000 legal immigrants came north across the border. The main causes of this increase were economic demand in the southwest for additional labor, and the decline in arrivals of Oriental labor into the Pacific region. This influx of Mexican immigrants was sharply curtailed to only 28,000 Mexican arrivals in the 1930s because of the Depression (Grebler et al., 1970; Samora, 1971). Owing to increased demand for cheap labor in the 1940s, temporary contract arrangements were conceived which became known as the Bracero Program. However, legal immigration remained very low (54,000) in the 1940s. In the period 1947–1949, legal agreements were enacted in the U.S. to formalize the contract labor arrangement, with the consequence that contract arrangements increased from 100,000 per year in 1950 to 400,000 per year in the late 1950s. In all, from the start of the bracero idea in the mid-1940s through its demise in the mid-1960s, about 5 million contract laborers were brought into the U.S.

With increasing employment opportunities in the 1940s, IMA entries increased greatly, as did IMA apprehensions, for which data are available. IMA apprehensions peaked at 1,000,000 in 1952, decreased to a low of 30,000 in 1960, and increased enormously from 1965 to 1975. An estimate of apprehensions for the Chula Vista Border Patrol Sector in 1973 extrapolates to a 1976 national level of about 950,000 apprehensions (Dagodog, 1975).

There has thus been a sequential series of migrations. Legal migration brought in large numbers in the 1930s; illegal entries reached a peak in 1952; the Bracero Program peaked in the late 1960s; currently, illegal entries are at a maximum, having increased tremendously since 1960, while legal migration is moderate and steady. The influence of this erratic international migration history on the county appears to have been minimal. In the 1930 Census, 35.4% of respondents were classified as Mexicans. Table 3–16 compares relative percentage "Mexican" in 1930 with percentage "Spanish language" in 1970.

Assuming some comparability of definitions, the SA population of Imperial County has remained relatively stable, but it has doubled for California. Thus, although waves of various types of migrants have come north from Mexico (many, in fact, through the county's border station), they have not altered Imperial's ethnic balance significantly in 40 years. The importance of these migratory patterns, relative to energy development, will be discussed later.

Socioeconomic comparison of the SA with the anglo population in 1970 reveals differences in spatial location, income, education, oc-

TABLE 3-16. SA POPULATION PERCENTAGE FOR TWO COUNTIES AND THE STATE[a]

Variables	Imperial	Kern	California
(a) 1930 percent "Mexican"	35.4	8.6	6.4
(b) 1970 percent "Spanish language"	44.9	15.5	13.7
Ratio (a)/(b)	1.27	1.80	2.14

[a] U.S. Bureau of the Census (1931 and 1971).

cupation, fertility, and age structure. All such differences have been previously noted for other regions (Bradshaw, 1973 and 1976; Grebler et al., 1970; Uhlenberg, 1973). The major geographical trend is the dominance of SAs in Calexico. Another important factor is the very high (81%) proportion SA in Heber.

The SA population has a very high birth rate, combined with a very high out-migration of young persons. SA fertility is documented to have been the highest of any major ethnic group in the U.S. (Bradshaw, 1973; Grebler, 1970; Uhlenberg, 1973). In Imperial County in 1970, SA completed fertility (or total average lifetime births to 1,000 women, 35–39) was 4,945 for SAs compared to 3,319 for anglos, a difference of 49%. Recent trends in the 1970s have shown a convergence of SA fertility rates with rates for the general population. The extent of this convergence in Imperial County will be influential in affecting projections of ethnic composition and population totals.

The net result of such a history of high fertility and high out-migration of young adults is a SA age structure of extreme youthfulness. Table 3-17 gives the 1970 age structures for SAs and anglos. The sharp contrast is shown by average age, which is about 33 for anglos and 19 for SAs.

About 27% of the SA population is age 0–9, and another 27% age 10–19. The influence of this youthfulness is pervasive—and serves to reinforce disparities in income and occupation. Unless altered, it will surely influence geothermal energy labor factors.

For income and education, there is marked contrast between the two ethnic groups. Table 3-18 shows 1970 county income data. The ratio of average income for anglo families and unrelated individuals (adults) to that of SAs is 142%, and the income distributions are quite opposite. Similarly, educational differences are marked. Only about 47% of SAs over age 25 are grade school graduates, compared to 86%

TABLE 3–17. AGE STRUCTURE BY ETHNIC GROUP AND SEX FOR IMPERIAL COUNTY, 1970[a]

Age group	Male			Female		
	Number	Percent		Number	Percent	
Anglo						
0–5	1,256	7.3		1,161	6.7	
5–9	1,670	9.7	35.0	1,566	9.0	34.5
10–19	3,074	18.0		3,268	18.8	
20–39	3,898	22.0		4,234	24.3	
40–64	5,407	31.8	65.0	5,410	31.1	65.5
65+	1,839	10.7		1,763	10.1	
	17,144			17,402		
SA						
0–5	2,082	12.4		2,197	12.5	
5–9	2,536	15.1	56.4	2,584	14.7	53.6
10–19	4,814	28.9		4,625	26.4	
20–39	3,440	20.6		4,384	25.0	
40–64	3,052	18.2	43.6	3,132	17.9	46.4
65+	808	4.8		606	3.5	
	16,732			17,528		

[a]Source: U.S. Bureau of the Census (1971).

of anglos. Similarly, for SAs over 25, 23% have high school degrees versus 55% for anglos. For college graduation, the respective rates are 2% and 10%. The extreme, youthful SA age distribution further increases the impact of these differentials.

In occupational distribution, the SA population shows an expected difference from anglos. Green-card border commuters were redistributed to the occupational categories in accordance with available statistics. An average number of 9000 (all classified males and SA) was assumed with 85% in farm occupations (U.S. Senate, 1971). With adjustments, the occupational contrast shown in Table 3–19 is striking—with 63% of the SA labor force being farm laborers and foremen, compared to 7.2% for anglos. By contrast, 39.2% of anglo males had white collar occupations (professional, managerial, sales, clerical) compared to 9.6% for SAs. The two female distributions are more similar, but the farm element still differs significantly (34% for SA, 9% for anglos).

As expected, in the major towns, there is greater tendency toward white collar occupations (43.4% for Brawley, 48.6% for El Centro, and 50.8% for Calexico, all figures unadjusted for border commuters). The figure for Calexico, combined with Calexico's large SA population, implies that SA white collar workers are concentrated there. SA unemployment and poverty rates are about twice those for anglos. SA poverty rate was 25.5% versus 10% for anglos and 12% for the state. Likewise, SA male unemployment was 9.6% versus 5.9% for anglos and 4.2% for California.

Illegal Mexican Aliens

The entry of illegal aliens into the U.S. has always been of great concern to the government. The Immigration and Naturalization Service has historically devoted most of its efforts to the control of illegal entrants. In recent years, the overwhelming majority of illegal entrants into the United States has been from Mexico.

To the extent that MAs are successful in entering and remaining in the United States and finding employment, the Mexican economy gains in two ways: persons are employed who would otherwise be a burden in their countries, and money is sent to Mexico from the United States by persons who would otherwise be unemployed.

TABLE 3–18. INCOME FOR FAMILIES AND UNRELATED INDIVIDUALS BY ETHNIC GROUP[a]

Income Class	Anglo		SA		Ratio of % Anglo/% SA
	Number	Percent	Number	Percent	
0–1,000	283	2.7	230	3.4	79
1,000–2,999	740	7.0	737	10.9	64
3,000–4,999	1,109	10.6	1,284	19.0	56
5,000–6,999	1,245	11.9	1,391	20.7	57
7,000–8,999	1,458	13.9	1,098	16.2	86
9,000–11,999	1,952	18.5	1,010	14.9	124
12,000–14,999	1,497	14.3	538	7.9	181
15,000–24,999	1,735	16.5	367	5.4	306
25,000–49,999	419	4.0	102	1.5	267
50,000+	58	0.6	8	0.1	600
	Anglo	SA	Combined	Income ratio Anglo/SA	
Number	10,496	6,765	17,261	1.42	
Average income	10,842	7,613	9,577		

[a] Source: U.S. Bureau of the Census (1971).

TABLE 3–19. OCCUPATIONAL DISTRIBUTION (IN PERCENT) OF IMPERIAL COUNTY POPULATION BY ETHNIC GROUP AND SEX FOR PERSONS AGE 16+ [a]

Occupation	SA	SA (adjusted) [b]	Anglo
Male			
Professional	4.6	1.8	11.4
Managers	8.7	3.4	16.5
Sales	5.3	2.4	6.0
Clerical	3.8	2.0	5.3
Craftsmen	12.4	6.6	20.7
Laborers and operatives	26.2	14.5	17.4
Farmers and farm managers	0.6	0.3	7.3
Farm laborers and foremen	30.2	63.4	7.2
Service	7.9	5.8	8.1
Private household	0.0	0.0	0.0
Total number:	5,865	14,865	14,912
Female:			
Professional	5.7		18.1
Managers	3.3		4.4
Sales	12.4		6.8
Clerical	27.7		39.4
Craftsmen	0.7		1.1
Laborers and operatives	13.5		4.1
Farm workers	14.0		1.4
Service	16.3		21.2
Private household	6.4		3.5
Total number:	2,702		5,380

[a] Source: U.S. Bureau of the Census (1971).
[b] Border commuters redistributed.

For the United States, the entrance of thousands of illegal aliens every year has many consequences. In the first place, the efforts to apprehend illegal entrants are costly. The annual budget for the Immigration and Naturalization Service in 1969 was $89,699,300. This included support for the Border Patrol, involving the services of several hundred officials and the maintenance of many offices, a training center, hundreds of vehicles including airplanes and boats, and costs of transporting aliens within the U.S. and abroad.

U.S. citizens who live in the border region and who lack sophisticated employment skills often look upon aliens as competition because they work for less and they are a readily available source of labor. SA citizens and Mexican citizens who are legal resident aliens in the United States often have to compare with the illegal aliens for the unskilled employment available in the border region. This is true in agriculture, industry, services, and domestic employment.

The illegal entry of such large numbers of persons creates a series of problems in communities where they live (Samora, 1971). Housing is perhaps the most critical of these. The IMAs have few choices as to where they will live and how much rent they will pay. The most primitive of accommodations are not unusual, particularly for those working in agriculture. Since the great majority of the aliens are men and in the younger age group, problems characteristic of homeless men are common to them, such as delinquency and crime. Mexican and United States agencies working in public health, welfare, police protection, narcotics, and many other fields have for years made joint efforts to solve these social problems.

Smugglers trafficking in human beings have made great profits from illegal aliens (Samora, 1971; Stoddard, 1976). The price for smuggling individuals is high, costs are relatively low, and penalties, if they are caught, until very recently, have been minimal and evidently well worth the risk.

The illegal movement of people across the Mexican border, which was supposed to have been controlled by 1956, reveals that there has been a steady increase from 1964 to the present time. Some persons, including Immigration and Naturalization Service officials, have suggested that the increase in IMAs is due to the elimination of the Bracero Program. There is some truth in this, but only up to a point. Even in 1954, when the Bracero Program was in full swing, over a million IMAs were apprehended—the highest figure in history.

The two most compelling reasons for illegal immigration are the insatiable demand for cheap labor in the United States and the tremendous population increase occurring in Mexico. Mexico's population growth rate is among the highest in the world. Although Mexico's economic growth programs have made enormous strides, the economy has not been able to provide sufficient employment, schools, income, and services for the increasing population. While its food production is high, the availability of food for the poor is a problem.

As agribusiness in the U.S. becomes larger, the situation encourages a great demand for a cheap and mobile labor force. The Bracero Program was just such a labor force and, additionally, it was partially subsidized by the federal government. IMA labor is even cheaper—no need for contracts, minimum wages, health benefits, housing transportation, etc. IMAs have few legal rights and can be dismissed without notice. A final reason for the increase in IMAs is that there are no great penalties involved in being an IMA nor in hiring an IMA. Both parties, employer and employee, can ignore legalities, the only serious consequence being an inconvenience to one party or both.

The above generalizations apply in various degrees to Imperial County. One fact is certain: the large number of IMAs present in the county. In 1975 there were approximately 50,000 apprehensions of IMAs in the county. This was determined by extrapolating earlier published figures in Samora (1971), using results of Dagodog (1975). One multiplier suggested by Samora for determining residents from one-year apprehensions is 3×. Since it is likely in the county's case that proportionately more IMAs are transients, one guess at IMA population would be to apply a multiplier of 1×, yielding a resident IMA population of 50,000.

The MA Population and Geothermal Energy Development

The county's MA population may influence potential geothermal power development and industrialization. Given the county's chronic out-migration, a key question becomes whether or not future geothermal employment will tap the existent SA labor force and/or offer inducement for young SA adults to remain in the county. If the above-delineated educational gap remains in effect, it will likely be more difficult for a potential SA to gain a job in geothermal energy development, compared to the average anglo. Unless the SA educational level increases, skill requirements for employees in geothermal power plants and multipurpose industries will be critical in determining the extent of employment benefits to a future SA labor force.

Perhaps the large size and sophistication of corporations developing the energy for multipurpose use will reduce potential benefits to SAs. If a governmental policy were developed to favor small businesses, perhaps more employment benefits would accrue to SAs. The union hiring practices of energy companies is also important. Most oil companies and Southern California utilities are union, while local companies and their employees tend to be nonunion. Also, SAs tend to be less union-oriented than anglos in Imperial County. Hence, hiring by outside energy developers will likely favor local or extralocal anglos over SAs.

In future population projections for the county by ethnicity, the key determinant will be the assumption of migration fluxes for anglos and SAs. It is too speculative to predict relative advantages and disadvantages for young adults to remain in the county 20 or more years in the future, when large-scale power plants and geothermal industries may be present.

Another unknown factor is the shape of future immigration laws and enforcement procedures and effectiveness. Judging by a volatile past history of regulatory shifts, there will be future changes, but these are impossible to predict, since they involve U.S. economic health, foreign policy, technological factors in agriculture, etc.

ACKNOWLEDGMENT

Parts of this chapter were taken from *Proceedings of the American Statistical Association*, 1976 Social Statistics section, pp. 666–672, by James B. Pick, Charles Starnes, Tae Hwan Jung, and Edgar W. Butler, and from the population section of the Final Report of NSF/ERDA grant No. AER 75-08793. Mike Pasqualetti, Mildred Pagelow, Peter Force, and Manuael Sanchez had made helpful suggestions in reviewing that paper.

4

Regional Employment Implications for Geothermal Energy Development in Imperial County, California

INTRODUCTION

This chapter examines possible employment and associated local labor market effects of geothermal development in Imperial County, California. We first discuss available historical evidence from similar past research and then examine studies relating to the prospect of geothermal development in Imperial County. Next, we examine U.S. Census statistics on the county by using statistical regression in order to understand its socioeconomic characteristics, and to project potential employment effects of geothermal development. Finally, we address the applicability and potential usefulness of regression results to prospective geothermal development.

As we pointed out in Chapter 3, Imperial County is a predominately agricultural region anticipating a rise in geothermal energy capacity, from 0 MW_e at present, to about 1500–3000 MW_e by the year 2000. Current local installed energy power sources are 150 MW_e hydroelectric and 350 MW_e fossil fuel. It is generally expected that beginning in the early 1980s there may be direct and indirect effects on county employment resulting from installation of geothermal power capacity, which will probably be added in 100–150 MW_e/year increments.

As in the previous chapter, data were obtained for Imperial County from the fifth count of the 1970 Census (U.S. Census, 1974). The Census divides Imperial County into 80 enumeration districts (EDs) which vary in size from 88 to 2261 persons. For the analysis in this chapter, the EDs are split into two subsamples—geothermal and nongeothermal. As in the previous chapter, known geothermal resource areas (KGRAs) have been delineated in the county. The detailed map shown earlier indicates the major towns, ED boundaries, and KGRA boundaries. Such a classification system results in 48 geothermal EDs and 32 nongeothermal EDs. There is a strong similarity in these categories to urban–rural categories: geothermal EDs are 87.5% urban and nongeothermal ones are 68.7% rural.

An important factor underlying Imperial County's economy is its independence of adjacent counties. It is difficult to study intercounty implications of employment for Imperial County. Total imports in Imperial County in input–output analyses are 41.4% agriculture related; foreign exports included in the gross output are 47.2% agriculture related; and other exports included in gross output of the county are 72.9% agriculture related (Lofting and Rose, 1976).

Imperial County's economy is not, in large part, tied to neighboring counties, but rather to a national and international agricultural distribution network governed by rather unpredictable commodity markets. Thus, the present analyses do not attempt to examine intercounty interactions. Many areas of the western U.S. contain large geothermal fields, but only the one in the Sonoma and Lake Counties, California, has been developed so far. Thus, the rationale in methodology, including ED data, may prove of benefit to other anticipated noncounty regional development locations.

An exceptional feature of Imperial County's economy is its independence of its neighbors in the U.S. and its economic and labor force association with Mexico. There are some complex border economic interactions, particularly in regard to the sister cities of Calexico in Imperial County and Mexicali, Mexico (Kjos, 1974). There is a large potential for cooperative economic exchanges at the border but so far attempts at such cooperation have not succeeded. The major type of border exchange that occurs currently is a highly seasonal population of Mexican citizens living in Mexicali who commute daily to primarily agricultural farm labor jobs in Imperial County. This largely male labor force is estimated to vary daily between 6,000 and 12,000, which translates to 20.3%–33.8% of the county's labor force. Unfortunately, no detailed socioeconomic information is available on these commuters or on other border economic interactions; thus, the only choice is to regard border effects as a source of error in considering the county as one labor pool.

In the present analysis, populations are determined by place of residence rather than by place of work. Place of work might be preferable if analyses were intended to emphasize characteristics of work-

Regional Employment Implications

ers as groups and businesses and industries and public sector work places. However, data on work place are unavailable at the subcounty level. Also, regional socioeconomic characteristics of unemployed persons are probably more meaningful in terms of similar residential locations. Also, in terms of application of this study to energy development, residential areas of the unemployed may be more important because businesses plan their prospective work forces at new locations in part on the basis of commuting distance for workers from their residences (Fulton, 1974).

The rationale behind the division of areas into geothermal and nongeothermal ones is the following: certain effects in geothermal energy development will be concentrated on persons in proximity to the geothermal fields. For example, possible hydrogen sulfide emissions, noises associated with geothermal power plants, danger from well blowouts, and nonelectrical uses of the energy are all localized. Nonelectrical use of the energy includes use of the hot water brought up to the surface for house heating, air conditioning, food processing, ground warming, extraction of chemicals, and baleanology (i.e., bathing and health spas). Since there is a technological limit on the transport of hot water for nonelectrical uses of perhaps 50 miles and a generally much shorter economic limit (see Chapter 8), geothermal EDs almost certainly will include land areas residentially closer to such uses.

Results of the regression analyses reported in this chapter may be cautiously utilized in assessing nonelectrical employment opportunities; they should be interpreted cautiously for the following reasons: (1) skill requirements for nonelectrical industry may override proximity in establishment of labor pools; (2) in future energy development, areas adjacent to geothermal fields may be altered by in-migration or out-migration, and (3) sampling in the regression analyses of only one annual and seasonal time point may not apply to later populations. In this chapter, a regression analysis of broad unemployment variables is done, using 1970 Census ED data.

A general caution on the regression approach is that these studies, done with 1970 data, precede the installation of any geothermal capacity, electrical or direct use. Hence, geothermal variables eligible for introduction into the regressions are limited in this study to variables for geothermal exploration, i.e., the KGRA boundaries. As geothermal development process unfolds, more meaningful regression studies will be possible.

The regression analysis investigates effects of socioeconomic characteristics of the unemployed. As shown in Table 4-1 (variable 12), the

TABLE 4-1. MEANS AND COEFFICIENTS OF VARIATION FOR INDEPENDENT VARIABLES[a]

Variables	Entire		Geothermal		Nongeothermal		California[b]
	Mean	Coefficients of variation	Mean	Coefficients of variation	Mean	Coefficients of variation	
1. Percent Spanish-American	0.455	0.606	0.541	0.568	0.323	0.456	0.137
2. Dep. ratio	0.957	0.258	0.952	0.291	0.964	0.205	0.606
3. White fertility ratio	0.465	0.391	0.457	0.339	0.477	0.456	0.378
4. Spanish fertility ratio	0.646	0.669	0.535	0.580	0.812	0.655	0.528
5. Percent now married	0.613	0.106	0.601	0.121	0.631	0.074	0.595
6. Percent different house 1965	0.549	0.250	0.547	0.246	0.552	0.260	0.434
7. Percent different county 1965	0.107	1.025	0.092	0.700	0.130	1.179	0.200
8. Percent different state 1965	0.075	1.485	0.084	1.441	0.062	1.551	0.092
9. Percent abroad 1965	0.051	2.115	0.061	2.207	0.035	0.955	0.025
10. Percent white collar	0.432	0.428	0.444	0.446	0.413	0.398	0.550
11. Percent blue collar	0.428	0.404	0.425	0.399	0.433	0.418	0.435
12. Percent farm worker	0.140	0.974	0.132	1.115	0.154	0.790	0.015
13. Density A[c]	0.577	0.266	0.547	0.264	0.623	0.254	0.672
14. Density B[d] (anglo)	0.807	0.137	0.794	0.161	0.826	0.093	0.920[e]
15. Density B (SA)	0.632	0.282	0.646	0.303	0.611	0.243	—
16. Percent utility gas	0.519	0.535	0.649	0.307	0.325	0.823	0.859
17. Percent bottled gas	0.098	1.597	0.041	1.656	0.182	1.135	0.031
18. Percent electricity	0.315	0.628	0.257	0.677	0.402	0.503	0.086
19. Percent rentals $0–$99	0.405	0.574	0.403	0.644	0.408	0.462	0.400
20. Percent rentals $100–$149	0.426	0.420	0.399	0.404	0.465	0.427	0.365
21. Percent rentals $150+	0.144	1.074	0.166	1.034	0.110	1.072	0.234
22. Percent income $0–$3.9[f]	0.194	0.751	0.189	0.754	0.201	0.756	0.265
23. Percent income $0–$9.9[f]	0.431	0.350	0.431	0.362	0.432	0.336	0.328
24. Percent income $10–$24.9[f]	0.372	0.597	0.376	0.636	0.365	0.539	0.407
25. Percent below poverty	0.169	0.803	0.180	0.773	0.152	0.858	0.084
26. Percent house value $0–$9.9[f]	0.290	0.855	0.252	0.873	0.348	0.803	0.047
27. Percent house value $25+[f]	0.074	1.779	0.081	1.799	0.063	1.712	0.412
28. Percent spanish tenure	0.506	0.549	0.539	0.482	0.458	0.658	0.549[a]

29. Percent house built 1965–1970	0.131	1.165	0.137	1.346	0.122	0.721	0.135
30. Percent house built 1960–1965	0.129	1.015	0.125	0.982	0.136	1.066	0.178
31. Percent house built before 1965	0.727	0.339	0.737	0.351	0.711	0.323	0.687
32. Percent elementary education	0.299	0.711	0.308	0.736	0.285	0.674	0.258
33. Percent H.S. education	0.469	0.323	0.443	0.328	0.508	0.305	0.549

[a]Sources: U.S. Census (1973, 1974).
[b]Statewide crude statistics from U.S. Census (1973).
[c]Three or less persons per unit.
[d]One or less person per room.
[e]Density B for the entire population.
[f]In thousands.
[g]Tenure for the entire population.

TABLE 4-2. MEANS AND COEFFICIENTS OF VARIATION FOR DEPENDENT VARIABLES[a]

	Entire		Geothermal		Nongeothermal		California[b]
Variables	Mean	Coefficients of variation	Mean	Coefficients of variation	Mean	Coefficients of variation	
34. Age eligibility (male)	0.640	0.171	0.621	0.148	0.669	0.189	0.693
35. Age eligibility (female)	0.651	0.143	0.633	0.117	0.677	0.166	0.713
36. Age eligibility (SA)	0.553	0.188	0.560	0.175	0.544	0.172	0.604
37. Age eligibility (white)	0.645	0.150	0.624	0.123	0.675	0.209	0.709
38. Labor force participation (male)	0.721	0.228	0.760	0.150	0.661	0.315	0.767
39. Labor force participation (female)	0.358	0.380	0.373	0.341	0.335	0.439	0.421
40. Labor force participation (SA)	0.495	0.401	0.546	0.235	0.418	0.255	0.583
41. Labor force participation (white)	0.532	0.246	0.555	0.196	0.497	0.310	0.567
42. Unemployment (male)	0.031	0.897	0.079	0.724	0.084	1.097	0.043
43. Unemployment (female)	0.061	1.074	0.058	0.918	0.066	1.240	0.029
44. Unemployment (SA)	0.071	1.035	0.061	1.283	0.077	0.903	0.046
45. Unemployment (white)	0.072	0.802	0.071	0.701	0.075	0.930	0.034

[a]Sources: U.S. Census (1973, 1974).
[b]Statewide crude statistics from U.S. Census (1973).

population sampled is occupationally only 14.0% farm laborers on the average, since border commuters are excluded. Hence, results of the regressions apply mostly to white collar and blue collar occupations. Regression analysis reveals greater subtlety in employment trends than the farm labor analysis. Implications of the regressions are applied to several aspects of prospective geothermal development, including nonelectrical employment.

All definitions of variables which are unusual or defined in several ways in the literature are clarified below: (I) white collar (variable 10) comprises the occupational categories professional workers, farmers and farm managers, other managers, clerical workers, and sales workers. Blue collar (11) comprises the categories craftsmen, operatives, and service workers. Rentals (19–21) is the proportion of number of renter-occupied monthly gross rents of a certain value to all renter-occupied units. Income (22–24) is for families and unrelated individuals 14 years and older. Poverty (25) is the proportion of families below poverty level to all families. Spanish-American tenure (28) refers to the ratio of Spanish American owner-occupied housing units to all Spanish-American-occupied housing units. Henceforth, Spanish-American is abbreviated as SA. Structural history of housing (29–31) is the ratio of year-round housing units built in certain periods to total year-round housing units. (II) Fertility ratios (3,4) refer to the ratio of persons under 5 years to females age 15–44 years. Mobility indices (6–9) refer to the proportion of 1970 population 5 years and older of a certain mobility state in 1965 to all persons 5 years and older in 1970. Education variables (32, 33) are proportions in the population 25+ years. (III) Percent SA (1) is defined as the ratio (persons of Spanish language or surname)/all persons. Dependency ratio (2) is defined as (persons under 18 plus persons 65 and over)/persons 18–64. Percent now married (5) is the ratio of presently married persons 14+ to total population 14+. The utility variables (16–18) refer to the proportion (particular fuel type in occupied housing units)/all occupied housing units.

The SA (Spanish-American) designation used in this chapter refers to the designation, Spanish language or surname discussed in Chapter 3. The SA designation is the broadest census designation for Spanish population. For detailed discussion of census rules for Spanish designations, the reader is referred to Hernandez *et al.* (1973).

The white designation used later on in this chapter refers to the entire county population minus Blacks and others (i.e., Indian, Japanese, Chinese, Filipino, and other). Unlike the anglo designation in the previous chapter, the white designation includes most of the SA population.

The anglo designation in Chapter 3 represented the entire population minus SAs. Hence, it includes Blacks and others. The reason for such a classification is that the category Blacks and others could not be subtracted from the entire population owing to lack of detailed age structure data on Blacks and others in census sources.

The inconsistency in Chapter 3 of classifying Blacks and others has a very slight effect on studies of Imperial County, because Blacks and others constituted only 7.6% of the 1970 county population. Blacks were 3.5% and others were 4.1% of the population. Even for the town or rural divisions studied, the one with the largest percent composition of Blacks and others in 1970 was El Centro at 9.8% so that the error would appear minor.

Independent variables were carefully selected from Table 4-1, using the following selection criteria: (1) variables of direct economic importance to unemployment (variables 10-12, 18-31 in Table 4-1), (2) intermediate variables found to have been of importance in prior U.S. regional population studies (such as Baer, 1972; Biggar and Butler, 1969; Collver, 1968; Galle et al., 1973; 1974; Heer et al., 1970; Levine, 1957; Pick, 1977; Sweet, 1972) such as density, fertility, and mobility variables (variables 2, 3, 6-9, 13-15, 32, 33), and (3) an ethnic variable (percent Spanish speaking), dependency ratio and utility variables (variables 1, 2, 16-18) which were of critical importance in previous analyses.

Twelve dependent variables (see explanation of term dependent below) were selected and are shown in Table 4-2. The rationale for choice of these is as follows. Imperial County has an unusual age structure. Forty-five percent of males and 44.3% of females are below age 20, compared to respective California figures of 38.2% and 35.6%. Also, the county historically has had high sex ratios (1.19 in 1950 and 1.24 in 1960), although these dropped to 0.98 in 1970. These sex ratios, combined with sharp out-migration in young adult age categories, have caused a highly irregular county population pyramid (see Figures 3-1 through 3-4 and accompanying text). Another irregularity is the very sharp difference between the SA and anglo pyramids: 54.9% of the SA population is under 20, compared to 34.7% for anglos. To assess the broad age-structural forces of unemployment, three related unemployment variables were chosen. The first, henceforth referred to as age eligibility (variables 34-37 in Table 4-2) is the ratio of population 16 years or older to total population, which indicates the segment of the total encompassed by the labor pool. The strict census definition of labor pool, which includes all aged persons, was adhered to in this study because employed aged persons are counted by the census. For the analysis, age eligibility was categorized separately by sex and by ethnic category. The reason for twice dividing the entire variable into two categories rather than dividing it into four subcategories, such as SA male, was to avoid statistical error from additional halving of sample units, and to allow broader comparisons by sex and ethnicity. As noted above, the categories white and SA are not mutually exclusive.

Labor force participation variables (38–41) are defined as the ratio of civilian labor force to all persons 16+. These introduce the small error of inclusion of military personnel in the denominator, but not in the numerator. This error is justified because the military cagegory is not a subject of analysis—the military is assumed independent of geothermal effects. The above exclusion allows the three sets of dependent variables to multiply to the ratio of unemployed to total population. A third set of dependent variables for unemployment (42–45) is defined in the standard manner as unemployed persons/labor force.

Major differences between geothermal and nongeothermal areas include lower percent SA in the nongeothermal region, higher SA fertility ratio, higher county mobility, lower state and international mobility, and a utility house heating mix with greater use of bottled gas and electricity (32% utility gas, 18% bottled gas, 40% electricity, and 10% other). This fertility difference, although it is for only two regions, corresponds to nationwide results showing a significant inverse relationship between nonwhite fertility and percent nonwhite (Pick, 1977).

Values for the dependent variables (Table 4-2) show lower age eligibility than for California—a reflection of the younger county pyramid, lower age eligibility for SA compared to anglos, labor force participation for males approximately twice that for females (a standard result—see Collver, 1968), lower county work participation than for the State, higher unemployment for males than for females, and for SAs than for anglos, and significantly higher unemployment for all categories than for the state. The latter difference, in general, may be due to a tighter labor market in the county, and is discussed in greater detail later with the regression results.

Regional Regression Analysis for Employment

Demographers, economists, and sociologists have for years advised that more industry should be located in areas of urban underemployment. They claim that new industries will increase the utilization of local labor supply, thereby reducing unemployment. Yet, examination of employment by industry in the varous labor markets shows a wide difference in the types of industries that are located in specific communities (Hunter, Reid, and Boddy, 1970).

This chapter section is concerned with local labor market analysis. Its main focus is on employment and associated local labor market effects of geothermal development in Imperial County. We first discuss available historical evidence from similar past research and then examine studies relating to the prospect of geothermal development in Imperial County. Next, we discuss regression findings of U.S. Census statistics on the county, in order to understand its socioeconomic characteristics, and to project potential employment effects of geothermal development. Finally, the applicability and potential usefulness of regression results to prospective geothermal development is addressed.

The present research does not examine total county employment shifts, but rather is an analysis of intracounty subregional (ED) differentials in employment. The basis in demographic theory for such a study may be found in regional studies of vital rates (Heer, 1970; Biggar and Butler, 1969; Melvin-Howe, 1968; DeSandre, 1971; Pick, 1977), overcrowding–isolation (Galle et al., 1974), female labor force participation (Collver and Langlois, 1962; Collver, 1968), male labor force participation (Baer, 1972), employment (Sweet, 1972), and others. These studies compare regions based on the methods of correlation and multivariate analysis. Regression analysis is one form of multivariate analysis used in the present geothermal study. Regression analysis is explained as follows (more detailed explanation is available in Draper and Smith, 1966). In a sample of a population, a variable y^* (dependent variable) is to be most closely predicted in an equation of the form below (linear) by a series of other variables, say, x_1, x_2, x_3, x_4 (independent variables):

$$y = ax_1 \pm bx_2 \pm cx_3 \pm dx_4 \pm e \qquad (1)$$

In order to determine the accuracy of estimated y from the equation in predicting the actual y^*, the difference $y - y^*$ is computed for each case in the sample and termed a residual. By the regression technique of least squares, the signs and absolute values of the constants a, b, c, d, e in equation (1) are varied until the sum of the squared residuals is minimized for the entire sample.

At this point, a least-squares regression equation has been selected. If the regression analysis has been performed with minimal violation of theoretical assumptions, the constants chosen will most accurately predict the ys for the sample population.

Regression analysis has the weaknesses of sampling inaccuracies, spurious correlation, nonorthogonality of independent variables, a normality assumption for the distribution of residuals, etc. However, results of such studies are useful if treated cautiously, with each of the above drawbacks in mind.

A design close to that of the present study was employed in a regression analysis of fertility (Heer and Boynton, 1970). The sample consisted of 591 U.S. counties in 1960, with a sampling ratio of 0.188—close to that in the present research. Baer (1972) performed a stepwise regression analysis of seven age-specific male labor force participation rates. These dependent variables were each defined as the number of males in the civilian age-specific labor force over the male noninstitutional age-specific population. Independent variables consisted of a mixture of age-aggregated and age-specific measures, including age-specific migration, age-specific percent in the military, ratio of age group to total population, median earnings of labor force males, education of males 25+ years, and male unemployment for the whole labor force. For most age categories, significant positive relationships were obtained for education and earnings, and inverse results were obtained for unemployment—the latter validating the "discouraged worker theory" in which worker discouragement in the face of high unemployment is postulated to reduce participation in the work force.

For sizable geographical areas (above, say, average U.S. county size) female labor force participation has been shown to have been nearly always lower than for males, most likely because of childbearing and family alternatives for women. In an international comparison labor force participation (defined as the ratio of number of people in the labor force to number of people 15–64) was 0.48 for France in 1954, 0.37 for the U.S. in 1950, 0.54 for Jamaica in 1953, and 0.12 for Egypt in 1947 (Collver and Langlois, 1962). Since that study, these figures have risen in many countries. For comparison, female participation, as defined above, is 0.48 for California in 1970 and 0.39 for Imperial County in 1970.

In a regional multivariate study, Sweet (1972) used the 1/1000 sample from the 1960 U.S. Census to study differences in employment of rural farm wives. He used the dummy regression method of multiple classification analysis, revealing effects on rural farm wife employment for such variables as southern location (positive), metropolitan county (positive), age range 20–39 (positive), Spanish-American (negative), elementary school education (negative), high income (negative), young children under 5 (negative), blue collar occupation (positive), and wage and salary income only (negative). All the major results corresponded closely to results for urban women.

Regional Employment Implications

FIGURE 4-1. Drill crew worker, Cerro Prieto, 1978.

There are few studies relating to application of employment statistics to rural industrialization. One study examined the following important question that also could be raised in connection with employment effects of industrial change—"When a plant actually does locate in a rural area, who will be employed?" (Gray, 1969).

When geothermal industries are established in Imperial County, employment opportunities may not open up for local workers. If the industry is capital intensive, employment may be available only for skilled workers. Drilling is a skilled job requiring extensive training, Figure 4-1. If local unemployed workers do not possess requisite job skills—which is likely—there will be few employment opportunities available. This would mean, of course, that industry would bring in workers from outside areas, thus leaving the local unemployed unaffected. Although additional peripheral jobs are created by the location of an industry, they may well be in the service or professional categories.

This point is substantiated by Gray (1969). In 1956, an aluminum reduction and rolling mill was built by the Kaiser Aluminum and Chemical Company at Ravenswood, West Virginia, a small town located in a depressed agricultural area. A decline in agricultural activities had caused a general out-migration from the area. As the Kaiser plant went into operation, an effort was made to give first hiring priority to workers from the Ravenswood area. This proved unsuccessful because local workers did not possess requisite job skills. Even though standards were relaxed, the qualifications of most of the unemployed and underemployed workers fell far short of those necessary to obtain employment; most workers had to be imported from areas outside the state. As the plant increased its operation, skill requirements also increased, and unskilled workers were gradually replaced. Although some 4000 workers were employed at the plant, only 300–500 of them came from the local area. Some of these had moved from the area, had acquired skills, and had returned home when the plant started operations. Moreover, the secondary effects on employment were also disappointing. Although a number of additional jobs were created in Ravenswood, those that required any degree of skill also were filled by outsiders.

A rather similar result was noted in a study of geothermal development on requirements for direct and indirect employment in Imperial County (Rose, 1977). Rose calculated the direct employment effect for all phases of a 50-MW_e development. These phases, with the number of required direct employees, both permanent and temporary residents, in parentheses, are exploration (15.2), field development (46.8), power plant construction (129.9), field operation (0.4), and power plant operation (5.2).

Of the total of 197.5 direct employees required, 76.2 were projected to be permanent residents. Likewise, the 50-MW_e development was estimated to result in 16.5 indirect and induced resident employees. On the scale of a geothermal development process to an eventual 2,000 MW_e capacity, such figures would imply 3,048 direct, resident employees and 660 indirect and induced, resident employees. These numbers are small relative to the 1970 county employment base of 23,479 workers. The above figures do not include an estimated 4,852 nonresident, temporary geothermal workers. Rose attributes such relatively small overall employment effects to "leakages in the economy and the absence of specialized goods, services, and personnel required in the exploration, drilling, and construction phases of geothermal development." Such figures might be increased by up to a factor of 5, if new industries, especially construction-related ones, were attracted into the county permanently.

These employment results are not contradictory to the population projections to be presented in Chapter 5, as the population projections encompass population increase from all causes, geothermal as well as nongeothermal.

Analysis Results

Correlation and stepwise regression analysis was performed by a standard routine BMDP-02R (Dixon, 1976). Values for the independent variables (Table 4-1) reveal most of the important socioeconomic features of the county covered in Chapter 3. Perhaps the most important feature is the large SA segment of the population—45% for the average ED. Overall county fertility is 22% higher than for the state for both ethnic populations and is 38.9% higher for SA than for whites. Other variables significantly different from statewide trends are mobility, which is higher for household and international and lower for county and state; utilities, which are higher for electrical and lower for gas heating; and poverty and housing, which reveal greater poverty and lower housing values than for California.

Correlations among independent variables for the entire sample reveal several pivotal variables highly correlated with others—percent SA, percent white collar, and white room density of 1.0 or less. There is a positive correlation ($r = 0.45$) between SA and dependency ratio—reflecting higher SA fertility. Percent SA has a negative relationship ($r = -0.61$) with percent now married, a result also of the significant younger SA age structure. SA predominance in farm labor category is reflected by a correlation of the two of 0.48.

SA population and income levels are strongly inversely related [r(% SA, % below poverty) $= 0.62$ and r(% SA, % income \$10,000–\$24,000) $= -0.51$], and the same type of effect holds for education [r(% SA, % elementary school education) $= 0.62$, r(% SA, % high school education) $= -0.64$]. All the above SA correlations remain strong, but reversed in sign if percent white collar or percent room density 1.0 or less is substituted for percent SA. This reversal is not surprising for percent white collar since Imperial County has an occupational structure with proportionately more anglos in the higher categories. However, the reversal with anglo density is surprising and is surely due to more than mere economics of house size. Density variables are important for the age eligibility regression discussed below.

The groups of independent variables show strong positive group intercorrelations; except for unemployment, there is a strong inverse correlation ($r = -0.38$) between male and female unemployment—a result of different social forces, as revealed by highly different regression results for unemployment by sex. For males, there are directional reversals on correlations for the three dependent variables(r (age eligibility, participation) $= -0.50$, r (participation, unemployment) $= -0.37$, and r (age eligibility, unemployment) $= 0.52$. This sequence underscores the difficulty in analyzing broadly unemployment relative to age structure—favorable age eligibility is related to poor work force participation, which in turn is related to high unemployment. The above pattern also holds to a lesser extent for anglos.

Results of the regression analysis for the three samples are presented in Tables 4-3 through 4-5. Dependent variables appear at the top of the tables and the independent variables appear on the left. The constants (Beta coefficients) in these tables have been standardized, meaning that all independent variables are scaled to vary between 0 and 1. For the entire county, age eligibility is strongly (positively) related to housing densities and strongly inversely related to fertility. This effect reflects a composite of demographic forces in the county— high fertility, a resultant large population proportion of children and adolescents, and consequent high housing densities. The inverse effect of fertility for females and SAs (versus little or no effect for males and whites) may reflect lower ED mobility, resulting in more stable ED fertility characteristics. Nineteen-seventy SA household and county mobility values are 0.40 and 0.08, versus 0.45 and 0.22 for non-SAs (U.S. Bureau of the Census, 1973). The argument that stability of residential location leads to more accurate measurement of period fertility (and mortality also) was used to justify a national study of vital rates (Pick, 1977). This residential stability is even more necessary in the present case for fertility to cause age eligibility changes, since to do so, the following line of reasoning is necessary: (1) present fertility in an ED is assumed to be highly correlated with the past birth sequence in an ED; (2) low mobility implies that the fewer babies born survived as ED residents; and (3) in a system more closed to migration, fewer past births result in an older age structure.

For the geothermal sample, the above conclusions again hold, with the addition of several economic results. For males, high income has a significant positive effect on age eligibility. Imperial County is estimated to have had an average annual net out-migration rate of 0.7% for 1930–1970. Thus, this income effect may be due to a lesser net out-migration of males from high-income EDs, relative to lower-income EDs—perhaps there are advantages for higher-income males to remain in the county. Nongeothermal sample results for age eligibility differ from those for the entire sample only for SAs. For SAs, the additional independent variable of county mobility has a negative effect, due perhaps to added fertility from young (18–mid-30s) county in-migrant parents combined with infants accompanying them.

Labor force participation for the whole sample has a significant inverse relationship with low income and poverty, and for females and whites, a positive relationship with high income. These results correspond closely to those of Baer (1972) for all male age categories except 14–19 and 65+.

Since 1970, this effect may have diminished, owing to the tendency for low-income workers who have lost their jobs to remain in the labor force (i.e., because they are looking for work, they are categorized as in the work force), in order to obtain unemployment benefits. For the geothermal sample, the above income effects and interpretations hold with three additional specific effects. For females, there is a positive relationship with percent electrical house heating. This is probably the result of urban gradients of significantly greater electrical usage in the higher-income areas of the three major towns. These gradients are reflected in correlations of electrical heating with income $0–$3,999, income $4,000–$9,999 and income $10,000–$24,999 of -0.49, 0.35, and 0.512. A related effect is a positive effect for males and whites from recent (1965–1970) house construction. This also reflects an income effect, as the above respective income correlations for housing construction 1965–1970 are -0.28, -0.37, and 0.42.

Also, in geothermal areas, there is a positive relationship, except for males, of labor force participation with high school education. This result also corresponds to consistent positive correlations of age-specific labor force participation with age-specific education (median school years completed) in the Baer (1972) study. Although there are no such present results for males, the arguments of Baer, emphasizing the importance of education to employers in job hiring, are applicable.

For the nongeothermal sample, the above effects for income (and highly correlated low rental variable) are interpreted as before. The significant inverse effect from interstate migration for males and whites is perhaps due to aged interstate in-migrants—mostly retirees—selectively locating in more rural areas of the county. This immigration of older residents is substantial—estimates of net migration 1960–1970 for males and females aged 65+ are 9.7% and 9.2%, respectively.

Not surprisingly, results for *employment* for the whole sample support, in general, a standard hypothesis that poverty and low income are associated with unemployment. For males, the most significant effect is a positive one with poverty. The significant positive effect from greater housing unit densities of 3 persons or less may be due to pos-

TABLE 4–3. STANDARDIZED BETA COEFFICIENTS, MULTIPLE CORRELATION COEFFICIENTS, AND PERCENT OF VARIANCE EXPLAINED BY SELECTED ENUMERATION DISTRICT CHARACTERISTICS FOR DEPENDENT VARIABLES, GEOTHERMAL SAMPLE[a]

Variables	Age eligibility				Labor force participation				Unemployment			
	Male	Female	White	SA	Male	Female	White	SA	Male	Female	White	SA
Percent SA												
Dependent ratio												
White fertility ratio			-0.299²						-0.280	0.428²		
SA fertility ratio				-0.442⁴					0.309¹	0.347²	0.511⁴	
Percent now married										0.303²	0.300	-0.442²
Percent different house 1965		-0.312²										
Percent different county 1965												
Percent different state 1965												
Percent Abroad 1965												
Percent white collar												
Percent blue collar					0.292¹							
Percent farm worker	0.486⁴											
Density A[c]	0.668⁴	0.760⁴	0.772⁴	0.523⁴								
Density B[d] (Anglo)				-0.461²								
Density B (SA)				0.480⁴								
Percent utility gas												
Percent bottled gas		0.206	-0.228¹									
Percent electricity							0.481⁴		-0.662⁴			
Percent rentals $0–$99										0.398²		
Percent rentals $100–$149										0.540⁴		

Percent rentals $150+								−0.481²				
Percent income $0–$3.9[e]		−0.412⁴	−0.270²									
Percent income $4–$9.9[e]												
Percent income $10–$24.9[e]	0.528⁴										0.349¹	
Percent below poverty					−0.453⁴		−0.692⁴					
Percent house value $0–$9.9[e]												
Percent house value $25+[e]												
Percent SA tenure				0.264	0.258²		0.271²					
Percent house built 1965–1970									−0.203			
Percent house built 1960–1965												
Percent house built before 1965							0.370					
Percent elementary education						0.283	0.453²		0.243			
Percent H.S. education												
Multiple correlation coefficient	0.790⁴	0.801⁴	0.832⁴	0.829⁴	0.820⁴	0.701⁴	0.836⁴	0.810⁴	0.534³	0.452²	0.681⁴	0.635⁴
Percent of variance explained (R^2)	0.623	0.641	0.692	0.687	0.672	0.492	0.670	0.656	0.285	0.204	0.463	0.403

[a] This table (and Tables 4-4 and 4-5) displays a series of 12 vertically oriented regression equations. For a simple exploration of regression, see text. The following is a simplified explanation of the table. The dependent variables (one for each equation) appear at the top of the table and the independent variables appear to the left. The beta coefficients (i.e., constants in the regression equation), appearing vertically under each dependent variable, have been standardized to vary between 0 and 1. In general, a larger absolute value indicates a stronger effect on the dependent variable. A positive sign for the coefficient implies a positive effect; a negative sign implies a negative effect. The percent of variance explained, appearing at the bottom of the table, gives the percent of variance (i.e., variation about the mean value) in the dependent variable which is explained by the selected independent variables. The multiple correlation coefficient is the square root of the percent of variance explained. The level of significance indicates the probability of exceeding particular values by random chance. For example, a significance level of 0.0005 indicates a 1 in 2000 probability of exceeding a designated coefficient by random chance.

[b] Levels of significance: ¹= 0.025, ²= 0.01, ³= 0.001, ⁴= 0.0005.

[c] 3 or less persons per unit.

[d] 1 or less person per room.

[e] In thousands.

TABLE 4-4. STANDARDIZED BETA COEFFICIENTS, MULTIPLE CORRELATION COEFFICIENTS, AND PERCENT OF VARIANCE EXPLAINED BY SELECTED ENUMERATION DISTRICT CHARACTERISTICS FOR DEPENDENT VARIABLES, ENTIRE SAMPLE

Variables	Age eligibility				Labor Force participation				Unemployment			
	Male	Female	White	SA	Male	Female	White	SA	Male	Female	White	SA
Percent SA					0.212							
Dependent ratio		$-0.303^{4,a}$	-0.155							-0.346^2		
White fertility ratio												0.362^3
SA fertility ratio				-0.564^4								
Percent now married												
Percent different house 1965					-0.206^1							
Percent different county 1965												
Percent different state 1965					-0.189							
Percent abroad 1965							-0.183^1					
Percent white collar			0.187^1									
Percent blue collar			0.629^4									
Percent farm worker			0.286^2									
Density A^b	0.795^4	0.395^4										
Density B^c (anglo)				0.412^4								
Density B (SA)				0.266^2								
Percent utility gas						0.223^2			0.350^4			
Percent bottled gas											0.256^1	
Percent electricity									-0.211			-0.242^1

Variable	(1)	(2)	(3)	(4)	(5)	(6)	(7)	(8)	(9)	(10)	(11)	(12)
Percent rentals $0-$99									0.290²			
Percent rentals $100-$149												
Percent rentals $150+												
Percent income $0-$3.9[d]			0.223¹									
Percent income $4-$9.9[d]		-0.712⁴	-0.263²									
Percent income $10-$24.9[d]					0.506⁴	-0.449³						
Percent below poverty						-0.413⁴	-0.658⁴					
Percent house value $0-$9.9[d]						0.216²						
Percent house value $25+[d]							-0.364					
Percent SA tenure								0.477⁴	-0.234¹	0.362³		
Percent house built 1965-1970			0.167									
Percent house built 1960-1965							-0.422²					
Percent house built before 1965												
Percent elementary education					0.294¹							
Percent H.S. education												
Female age eligibility[e]								-0.476⁴				
Multiple correlation coefficient	0.775⁴	0.760⁴	0.834⁴	0.732⁴	0.734⁴	0.738⁴	0.753⁴	0.501²	0.617⁴	0.464⁴	0.490⁴	0.441⁴
Percent of variance explained (R^2)	0.601	0.577	0.695	0.530	0.539	0.545	0.567	0.251	0.381	0.215	0.240	0.195

[a] Levels of significance: 1 = 0.025, 2 = 0.01, 3 = 0.001, 4 = 0.0005.
[b] 3 or less persons per unit.
[c] 1 or less person per room.
[d] In thousands.
[e] Only included for female unemployment.

TABLE 4–5. STANDARDIZED BETA COEFFICIENTS, MULTIPLE CORRELATION COEFFICIENTS, AND PERCENT OF VARIANCE EXPLAINED BY SELECTED ENUMERATION DISTRICT CHARACTERISTICS FOR DEPENDENT VARIABLES, NONGEOTHERMAL SAMPLE

Variables	Age eligibility				Labor force participation				Unemployment			
	Male	Female	White	SA	Male	Female	White	SA	Male	Female	White	SA
Percent SA												
Dependent ratio												
White fertility ratio												
SA fertility ratio				-0.570^{4a}								
Percent now married												
Percent different house 1965				-0.483^{3}								
Percent different county 1965												
Percent different state 1965					-0.488^{4}		-0.411^{4}	-0.408^{2}				
Percent abroad 1965										0.326		
Percent white collar												
Percent blue collar							-0.278^{2}					
Percent farm worker				$.507^{4}$								
Density A^{b}	0.688^{4}	0.603^{4}	0.491^{4}									
Density B^{c} (Anglo)		0.370^{2}	0.472^{4}									
Density B (SA)											-0.577^{2}	
Percent utility gas												0.478^{2}

Percent bottled gas				0.474[4]								
Percent electricity									−0.366[2]			
Percent rentals \$0–\$99								−0.478[2]			0.284[1]	
Percent rentals \$100–\$149												
Percent rentals \$150+												
Percent income 0–3.9[d]					−0.667[4]							
Percent income 4–9.9[d]												
Percent income 10–24.9[d]			−0.258[2]			0.784[4]	0.647[4]		−0.461[2]			
Percent below poverty											0.845[4]	
Percent house value 0–9.9[d]									−0.634[4]		−0.740[4]	
Percent house value 25+												0.467[2]
Percent SA tenure												
Percent house built 1965–1970												
Percent house built 1960–1965								−0.723[4]				
Percent house built before 1965								−0.455[1]		−0.440[2]		
Percent elementary education			0.220[1]						0.573[4]			
Percent H.S. education												
Multiple correlation coefficient	0.858[4]	0.835[4]	0.931[4]	0.835[4]	0.833[4]	0.784[4]	0.871[4]	0.743[4]	0.826[4]	0.548[2]	0.821[4]	0.637[2]
Percent of variance explained (R^2)	0.736	0.697	0.867	0.698	0.695	0.615	0.758	0.552	0.683	0.301	0.674	0.406

[a] Levels of significance: [1] = 0.025, [2] = 0.01, [3] = 0.001, [4] = 0.0005.
[b] 3 or less persons per unit.
[c] 1 or less person per room.
[d] In thousands.

itive effects on employment from the higher housing unit densities associated with families—family responsibilities are associated with greater employment. Since regression analysis for female unemployment against the independent variables was nonsignificant for the entire sample, it was decided, in this single instance, to include the non-unemployment-dependent variables as independent variables. The resultant inverse effect of female age eligibility may well be due to a positive effect with fertility [r (female age eligibility, white fertility) = −0.41; r (female age eligibility, SA fertility) = −0.13]. Sweet (1972) demonstrated a positive effect on unemployment of young children for rural farm wives, corroborating his own results for urban wives. The positive unemployment effect of rentals $0–$99 corresponds to the basic unemployment hypothesis of this paper but is contrary to Sweet's results for rural farm wives, as well as to results cited by Sweet on urban wives. The inverse dependency ratio effect would appear contrary to the age eligibility effect, and may be due to a regression error. For whites, the positive effect of poverty corresponds to the explanation given above for males. The positive effect of bottled gas is likely due to the greater presence of bottled gas for house heating in poorer areas (r bottled gas, income $0–$3999 = 0.31).

SA unemployment is most strongly (positively) related to the white fertility ratio. This is likely the result of high SA fertility [r (SA fertility, white fertility) = 0.36], with an explanation corresponding to that of Sweet referred to before in the discussion of female unemployment. The negative effect from electrical heating likely follows the same reasoning for electrical heating given above for female geothermal work participation. Unemployment shows greater differences between geothermal and nongeothermal than did the previous sets of dependent variables. For geothermal areas, all categories have positive relationships with fertility—as explained above. However, the reason for the greater importance of fertility in geothermal, as opposed to nongeothermal, areas is unexplained. The positive relationships with poverty for males and with gas heating usage for anglos are direct economic effects—gas usage follows a pattern opposite to that for electricity.

For the nongeothermal region, economic-related variables and migration variables are dominant for the non-SA categories. These are mostly self-evident. For example, for women, the percent below poverty exerts a positive effect on unemployment. One unexplained effect, for males and anglos, is the positive employment effect of low house values. This appears contradictory and may be spurious.

For nongeothermal SAs, the positive effect of percent married on employment is ascribed to marital pressures for economic support. The lack of such an effect in the geothermal (more urban) areas may possibly be due to increased welfare support and a larger job market easing the risks of unemployment. There is a significant negative employment effect from household mobility, the result perhaps of mobility-associated job disruption. Finally, the reduction in employment from increased percentage of owners of $25,000+ homes would appear contradictory, and perhaps is spurious.

IMPLICATIONS OF GEOTHERMAL DEVELOPMENT ON REGIONAL EMPLOYMENT

Imperial County has had a stable agricultural-dominated economy for decades. For these years, the county work force has had lower age and educational levels than the state. Subject to a 0.7% annual out-migration, the county has lost many of its skilled and trained youth—lowering the average skill levels of the county. Finally, there is a large, highly seasonal, and currently about half non-U.S.-resident farm labor work force commuting from Mexico. The prospective advent of geothermal energy development may alter each of the above fundamental county economic factors. The dominance of agriculture may be reduced or displaced by a large-scale energy-industrialization process. It is also necessary to consider concurrent trends, such as rising salinity of the Colorado River, virtually the sole source of agricultural water, which may reduce crop yields and cause further shifts to salt-tolerant but often less profitable crops.

The second fundamental feature, the presence of a large and partly nonresident farm labor segment may be somewhat altered by energy development. Based on geographical and agricultural data from Johnson *et al.*, (1976), estimates were made of farm labor displacement for the medium growth scenario with the capacity beginning in 1990 rising to a steady state level of 3000 MW$_e$ by the year 2010. (These estimates are covered in more detail in the next chapter.) Assuming a maximal agriculture land loss of 35% and well-siting areas from geothermal power plant and well acreage, transmission lines, access roads, possible land subsidence, etc., only a total of 150 farm laborers are estimated to be displaced from agriculture in 2010, assuming constant 1970 agricultural productivity. Hence, a substantial low-skill farm labor segment will likely remain present in Imperial County even after geothermal development reaches a stable state unless other factors result in a decline of county agriculture or large-scale industrialization utilizes extensive agricultural land.

The third major historical influence of persistent net out-migration, especially of young adults, may be altered by the attraction of jobs directly and indirectly created by geothermal energy. However, concurrent migration such as the recent national trend of net return to rural areas, may alter Imperial's migration streams as much as or more than geothermal development. Two factors mitigating against substantial in-migration are the county's inhospitable summer climate and lack of cultural attractions characteristic of the other major labor force areas of California. The latter attractions might appear, of course, as the result of energy development. Judging from the results of the Gray (1969) study, a large proportion of potential local geothermal workers probably do not possess the appropriate skills for newly created jobs, and thus may continue to out-migrate. This problem appears to be potentially severe for young SAs who possess below average education and training. Even in the current pregeothermal labor market, the high level of unemployment suggests a population pool of employables available, especially SAs who need to be trained for more skilled jobs. It would appear important for geothermal planners and policy makers to consider establishment of on-the-job training or training centers in order to avoid either continuing departure of lesser skilled young adults or the necessity for workers to travel outside the labor region to obtain retraining, as was done in West Virginia.

Having detailed several fundamental county economic forces and their likelihood for change, how do the results of these analyses help elucidate such future processes? One inability of regression is indication of cause and effect, which is further complicated by economic processes that are often circular. Nevertheless, one may speculate on several important age eligibility and participation implications as a result of geothermal energy development. The first of these are potential enhancement of age eligibility if future county fertility levels are reduced, potential increases in labor force participation if average county personal income levels rise from geothermal development, potential decreases in labor force participation if in-migration of aged persons for retirement increases, a distinct possibility if current trends continue. Regression implications for unemployment are more complex and must be examined by category. For male and white categories, a direct reduction in unemployment with more ample personal income and decreased poverty probably stemming from geothermal development may be inferred. For females, unemployment may be reduced through higher age eligibility, if county fertility levels are indirectly reduced by geothermal development. The model for this indirect reduction is hypothesized as follows: (1) geothermal development and related industrialization will increase the volume of in-migration into the county; (2) an increase will occur in the county's overall percentage anglo, owing to a higher percentage of anglos in the county's in-migration stream from (1). (This would stem from closer correspondence to statewide in-migration percentages); (3) a higher percentage of anglos will lead to a reduction in the county fertility rate. Finally, SA unemployment may be reduced if fertility lessens owing to the positive SA fertility correlation we already discussed.

In general, geothermal development influence on the county fertility pattern by an increasing urban proportion or by higher proportions of anglo in-migrants may result in an older age structure. This, combined with lower fertility and geothermal-development-related increases in personal income may effect increases from the present low levels of county employment, relative to total population, age-eligible population, and the work force.

A feature of geothermal development discussed earlier is geographic localized effects, such as certain pollutants and nonelectrical uses. If nonelectrical employment increases are even moderately localized in geothermal EDs, the regression differences between geothermal and nongeothermal areas may be of use to planners.

An example of such differences is the greater positive influence of fertility on unemployment in geothermal areas relative to nongeothermal areas. Therefore, the policy implications are that fertility reduction might contribute towards reducing unemployment in areas of nonelectrical industries (i.e., in geothermal areas). Such an inference is based on the assumption that geothermal energy development would not substantially alter the social forces on unemployment prevailing in the county in 1970.

An improved approach to policy formulation would be to update these regression studies periodically as new data become available during the geothermal development process. In such a case, not only would updated social data become available, but, also, more significant geothermal variables would be available besides the KGRA boundaries used in the present study. It should be reemphasized that the present study is *prospective*, since geothermal capacity, electrical or direct use, is just now being developed in the county.

In future studies, regressions could be performed based on geothermal data for power plant generating capacity, nonelectrical transmission flow rates, geothermal investment, number of geothermal employees, and so forth.

Acknowledgment

This chapter contains some material that was included in the *Los Angeles Council of Engineers Scientists Proceedings Series,* Vol. 3, 1977, pp. 159–167.

5

Projected Population, Growth, and Displacement from Geothermal Development

INTRODUCTION

With a complicated geothermal energy source, there are many pathways that the development process can take, depending on such factors as total amount of recoverable energy, land ownership, permitting and regulatory processes, drilling costs, community and extralocal leadership, energy consumer market area, etc. It is impossible to project all such unknowns ahead of time, in part because there is only one U.S. geothermal field in active production—the The Geysers steam resource with about 600 MW_e of installed electrical generating capacity. Hence, projections for different types for Imperial County can only be performed with simplifying assumptions. County population projections were done, based on differing assumptions of buildup in geothermal capacities (Pick et al., 1976). These, in turn, were used to project county interindustry interactions (Lofting, 1977) and county revenues and taxes (Rose, 1977).

In this chapter, reduction in the farm labor segment of the Imperial County labor force is projected based on losses in agricultural land directly caused by geothermal plants and wells. A 100-MW_e power plant and well siting area was assumed to spread over 650 acres of land. The proportion of land used in the well siting area by well pads, pipe lines, and access roads, by less predictable land subsidence, and by other factors were assumed at the levels of 5%, 10%, and 35%.

Three scenarios of future power plant capacity in agricultural county areas were assumed. Ratios of farm laborers to land area, based on studies of Johnson (1977) and Sheehan (1976), were then used to project geothermal farm laborer displacement. For 35% interstitial land reduction, $4,000 farm worker income, and the medium power plant scenario, the displacement is projected for year 2020 as only 1.96% of the 1970 farm laborer category.

In addition, this chapter contains a variety of population projections based upon a series of assumptions that are made explicit for each projection.

FARM LABOR FORCE REDUCTION BASED ON LAND AREA ANALYSIS

Table 5–1 presents the aggregated employment categories in Imperial County and the U.S. for the last three U.S. Censuses of Population. As expected for an agricultural county, the farming category is greatly enlarged relative to the U.S. as a whole. The 21% reduction in total percentage of the farm laborer category between 1960 and 1970 is exaggerated because of the presence of thousands of border commuters, mostly farm laborers, who live in Mexicali, Mexico, directly across the border, and commute to work daily in Imperial County. Such persons are not counted by the U.S. Census, since the Census counts persons based on residence (not work place) in the U.S. (U.S. Senate, 1971). These 1970 commuting workers were mostly residents of the county in 1960, prior to the end of the Bracero Program (Samora, 1971).

These data are important because they have been misinterpreted by several investigators of conditions in Imperial County. For example, Layton and Ermak (1976) remarked, and Ternes (1978:47) elaborated, upon the comment that "recent county employment patterns show a steady decline in agricultural employment (down 48% between 1960 and 1970) as agriculture has become mechanized." While agriculture may have become more mechanized during this period, the discrepancy in agricultural workers is primarily attributed to the census *not* counting Mexican national commuters in the labor force in 1960 but doing so is previous years (see Table 5–1). Thus, with commuters included in the analyses, there was a very slight *increase* in agricultural employment in Imperial County between 1960 and 1970.

The addition of the average of 6350 male commuting farm workers to the 1970 U.S. Census employment distribution (see Table 5–1) gives a 1970 farm worker percentage of 40.4%, quite similar to that of 36.9% in 1960, and a 1970 total of 9537 farm laborers. Such a large proportion of county employment in this category warrants the special projections in this chapter. It is important to note that this total is also

TABLE 5-1. Percent Males in Employment Categories for Imperial County and the United States, 1950–1970[a]

	United States			Imperial County			
Variables	1950	1960	1970	1950	1960	1970	1970[b]
Total male labor force (in thousands)	42,554	45,686	50,002	18.6	21.6	15.4	21.7
Aggregated employment category							
Farm	14.9	8.1	4.4	40.1	43.2	20.7	43.8
Farmers and farm managers	10.0	5.3	2.7	9.2	6.3	4.8	3.4
Farm laborers and foremen	4.9	2.8	1.7	30.9	36.9	15.9	40.4
Craftsmen and operatives	39.6	39.9	38.2	26.0	23.5	31.9	22.6
Clerical and sales	12.8	13.5	14.3	7.4	6.9	10.4	7.4
Professionals and managers	17.7	20.2	23.6	13.9	14.6	22.2	15.7
Other	15.0	18.3	19.5	12.6	11.8	14.8	10.5

[a] Source: U.S. Bureau of the Census: 1975, Historical Statistics of the United States; 1973, General Population Characteristics, Final Report PC(1)-B.
[b] Border commuters included.

affected seasonally by harvesting cycles. Data on the present analysis are based on the census date of April even though maximal county employment due to crop cycles is in January.

The reduction from geothermal development of farm laborers in the Imperial County labor force was estimated based on prior studies of geothermal capacity (Davis, 1976), crop acreage (Johnson, 1977), and power plant impact (Sheehan, 1976; Rose, 1977). This analysis is based on the following assumptions for land reduction at one power plant site.

It is assumed that a 100-MW$_e$ power plant installation will consume ten acres (i.e., remove ten acres from agricultural use by either direct or indirect effects) for the immediate area that the central power plant is sited on and the right-of-way for the central power plant. It is assumed that for the 100-MW$_e$ capacity, there are 20 production wells and 12 reinjection wells. Each well is assumed to be spaced over 20 acres. The total well spacing is thus assumed to consume 640 acres. These land areas are shown in Figure 5-1. The key question is, then, how much of the 640 acres is consumed either by well pads, pipe lines, right-of-way, subsidence problems, direct use applications of geothermal energy, and other environmental and agricultural causes. There are so many regulatory, agricultural, and geological unknowns in the above assumptions that the present analysis simplified matters by assuming three possible percentages of land reduction for the well-basin area (i.e., the 640 acres/100 MW$_e$ capacity): (1) 5%, (2) 10%, and (3) 35%. These reductions are henceforth referred to interstitial land reductions (ILR).

As in all other parts of this research utilizing the concept of KGRA, there is an error introduced by use of the geographic entity of a KGRA. The reason is that, presently, significant portions of an earlier designated KGRA are not hot enough to be productive, while areas outside earlier designated KGRA boundaries are hot enough to be commercially productive. The present KGRA boundaries for Imperial County were based on results presented by Rex (1971). For this reason, in place of KGRA, a much more accurate term would be *productive geothermal area*. One reason for not using this new term throughout the book is the convention of nearly universal use of the KGRA terminology. A second reason is that, although erroneous, the KGRA boundaries are available and the boundaries used for various analyses (including the U.S. Geological Survey). However, in the near decade since that time, there have been many additional discoveries and changes in the resource topography. Hence, it is likely that for the county as a whole about 40% of the area inside the county's KGRA

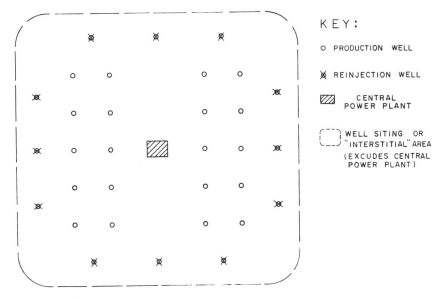

FIGURE 5-1. Land use for hypothetical 100-MW geothermal production field.

boundaries is noncommercial; an equivalent area outside the KGRA boundaries is capable of commercial development.

The present analysis, however, is not jeopardized by such an inaccuracy for the following reason: as a just approximation, the resource deletions are about equal to the resource additions. Furthermore, the resource additions are often near the KGRA, which has the deletion. For the second reason, all the other analyses in this book based on KGRA boundaries, such as the discriminant and regression analyses presented in Chapters 3 and 4, are not significantly affected by KGRA errors, because of compensating balances.

In the agricultural portion of the county, pressures for slant drilling will come from farmers. Slant drilling is drilling in a direction other than vertical. Farmers seek minimal disruption to their fields and wish drilling to be done only at the field edges. Drilling areas can be placed at the edges affecting perhaps only two or three rows of tile drains. (Tile drains are drains placed several feet underneath most of the county's agricultural fields to prevent salt from seeping upwards.) In slant drilling, a number of holes are drilled in various directions from a single well pad at the field edge. In this case, land is removed from agriculture by centralized well pads, pipe lines, and right-of-way, subsidence, and environmental courses. Thus, for the likely widespread use of slant drilling, except on the East Mesa, one could possibly choose a smaller percentage reduction based on the partial benefits to land alteration of this method. Nevertheless, some agricultural land would still be utilized.

For all three ILR scenarios, an important consideration is the probable extent of land subsidence; although not a part of this analysis, the East Mesa KGRA is not susceptible to geothermal subsidence due to the strength of underlying formations. In agricultural areas, such subsidence, whether caused by geothermal withdrawals or ground water pumping, would endanger farm lands because most of the canals supplying irrigation water to these lands are dependent on slight elevational inclines which could be altered by subsidence. The canals would not be easily adjustable because they are predominantly concrete. In other KGRA areas, there is only minimal danger of geothermal subsidence.

Another potential problem in the East Mesa and Heber KGRAs is subsidence due to the increasing use of ground water pumping in the Salton Trough south of the international border. The ground water aquifer underlying the Salton Trough deepens from the Imperial Valley portion, where it is about 6–8 feet to the Mexicali Valley portion, where some of the supply of the Mexican aquifer is from spillage of water from the All American Canal and it is as deep as 600 feet. Because the water flow within this aquifer is from north to south, pumping south of the border may cause land subsidence at shallow depths in the East Mesa and Heber KGRAs.

These three reductions correspond to generalized land use scenarios. The usual ILR is assumed to be from 2½% to 5%. Hence, the 5% ILR case presents the usual situation. The 10% ILR case is included for planning purposes as an upper range of ILR, with nonelectrical uses excluded. The case of 35% ILR is included as an upper range of labor force impacts if direct uses of geothermal become widespread in the well siting area. These uses may be industrial or agricultural, e.g., earth warming, hothouses, fish cultivation, drying, etc. For either type of direct use, the unskilled farm labor displaced is likely to be replaced by skilled industrial workers who will operate the geothermal power plants, Figure 5–2, or agricultural workers. The labor intensity of direct use applications may cause more skilled workers to be added than unskilled workers to be displaced. Since the computations performed and figures presented are for unskilled farm laborer category of the labor force, this ignores direct use additions to the labor force. However, since labor requirements of direct use have not been adequately

FIGURE 5-2. 50-MW turbine, Cerro Prieto. How many for Imperial County?

studied at existing direct use sites and since these requirements vary among the multitudinous kinds of direct uses, the present analysis would not be able, in any event, to estimate skilled, direct use workers added. This ignores direct use additions. In summary, the 35% is a generalized ILR case representing the effect of geothermal land use on the unskilled farm laborer category. These results should not be interpreted as estimating the effects on the general labor force.

This analysis differs from the East Mesa KGRA relative to the other KGRAs. The East Mesa KGRA lies on nonagricultural federal land to the east of the southeastern part of the agricultural valley. Estimates of potential are from 400 to 600 MW_e. It has the advantage of very low salinity and disadvantages of somewhat lower temperatures and costs of pumping water in and out. As subsequently discussed in Chapter 8, it differs from the other KGRAs by having a very large deposit of groundwater which might be used for cooling purposes.

Most important for the present analysis is the fact that the East Mesa KGRA will result in no farm labor displacement, since it has no agricultural production. Hence, it has been excluded from the present calculations.

Farm Labor Reduction

The calculations of farm labor reduction are shown in Tables 5–2 and 5–3. Two types of coverage are assumed for geothermal development—an average hundred acres and all field crops. Table 5-2 presents the computations for an average 100 acres for crops as defined by Sheehan (1976) and Rose (1977). For each KGRA, a manpower reduction (in man year units) is assumed based on the above analysis. Also, a scenario power plant capacity is assumed for each KGRA based on the medium estimate of Davis (1976). The KGRAs are then summed in the right-hand columns to give total farm labor displaced by interstitial land reduction, farm income, KGRA, and year for field crop areas. As shown in Table 5–2, the overall farm labor displacement ranges for all fields from a low level at 5% ILR of 27 or 34 laborers displaced to a high level at 35% ILR of 150 or 187 laborers displaced by the year 2020. Similar impacts are shown in Table 5–3.

The general county results are also shown in Table 5–4. The maximal reduction in labor force by the year 2020, assuming a 35% ILR, average crops, and $4,000 income, is 187 laborers, assuming fixed agricultural mechanization and other current trends. Based on the present labor force, of 29,829, including 6,350 border commuters, this is a reduction of less than 1% (0.62%). Since it is likely that the county will increase in population, this percentage may be reduced by 50% based on year 2020 populations. The 5% ILR for the above assumptions of unchanged yields results in a farm labor reduction of only 27 laborers, or 0.09% based on the 1970 population. For all field crops, the figures are roughly 75% less than for average crops. Based on a 1970 farm labor category, including border commuters, of 9,537, the 35% ILR and 5% ILR reductions are 1.96% and 0.28%, respectively. This analysis simplifies the impact of farm labor by extrapolating present electrical generation technology and agricultural land/labor ratios into the future. As discussed earlier, another source of error is lack of consideration of effects of direct use of geothermal energy on farm labor. For example, the number of farm workers per acre might be increased by agricultural direct use applications, such as hothouses, fish ponds, etc. These effects were not analyzed, because of lack of data.

Population is an important variable in energy studies. Fowler (1975) cites data showing that 19.2% of U.S. end use energy consumption in 1968 was residential and 24.9% was transportation-related. Both residential and transport consumption, to a significant degree, are functions of population. For neither of these uses does the popu-

TABLE 5–2. FARM LABOR FORCE DISPLACEMENT BY KGRA, PLANT CAPACITY, FARM WORKER INCOME, AND LAND REDUCTION FOR AVERAGE 100 ACRES[a]

Variables	Year	KGRA												Total	
		Salton Sea			Brawley			Heber							
		Capacity	FLD5	FLD4	Capacity	FLD5	FLD4	Capacity	FLD5	FLD4		Capacity	FLD5	FLD4	
5% ILR	1980	90	.51	.64	35	0.18	0.22	35	0.66	0.83		180	1.35	1.69	
	1990	250	1.42	1.77	100	0.51	0.64	100	1.89	2.36		500	4.77	5.97	
	2000	1250	7.16	8.95	500	2.55	3.19	500	9.47	11.84		2500	19.18	23.97	
	2010	1750	10.02	12.52	700	3.57	4.46	700	13.25	16.56		3500	26.84	33.55	
	2020	1750	10.02	12.52	700	3.57	4.46	700	13.25	16.56		3500	26.84	33.55	
10% ILR	1980	90	0.91	1.14	35	0.31	0.39	35	1.17	1.46		180	2.39	2.99	
	1990	250	2.52	3.15	100	0.90	1.12	100	3.33	4.17		500	6.75	8.44	
	2000	1250	12.62	15.77	500	4.49	5.62	500	16.68	20.85		2500	33.79	42.24	
	2010	1750	17.66	22.07	700	6.30	7.87	700	23.35	29.19		3500	47.31	59.14	
	2020	1750	17.66	22.07	700	6.30	7.87	700	23.35	29.19		3500	47.31	59.14	
35% ILR	1980	90	2.87	3.59	35	.99	1.24	35	3.69	4.61		180	7.55	9.44	
	1990	250	7.98	9.97	100	2.84	3.56	100	13.18	16.48		500	24.00	30.00	
	2000	1250	39.90	49.87	500	17.78	22.23	500	52.74	65.93		2500	110.42	138.02	
	2010	1750	55.85	69.82	700	19.92	24.90	700	73.84	92.30		3500	149.61	187.01	
	2020	1750	55.85	69.82	700	19.92	24.90	700	73.84	92.30		3500	149.61	187.01	

[a] ILR = interstitial land reduction; FLD5 = farm labor displacement (average annual income = 5,000); FLD4 = farm labor displacement (average annual income = 4,000); all capacity figures in MW.

TABLE 5-3. Farm Labor Force Displacement by KGRA, Power Plant Capacity, Farm Worker Income, and Land Reduction Percentage for Field Crops[a]

		KGRA											
		Salton Sea			Brawley			Heber			Total field crops		
Variables	Year	Capacity	FLD5	FLD4	Capacity	FLD5	FLD4	Capacity	FLD5	FLD4	Capacity	FLD5	FLD4
5% ILR	1980	90	0.23	0.29	35	0.07	0.08	35	0.07	0.08	180	0.37	0.46
	1990	250	0.64	0.80	100	0.19	0.24	100	0.19	0.24	500	1.02	1.27
	2000	1250	3.20	4.00	500	0.96	1.20	500	0.97	1.21	2500	5.13	6.41
	2010	1750	4.48	5.61	700	1.34	1.67	700	1.36	1.70	3500	7.18	8.97
	2020	1750	4.48	5.61	700	1.34	1.67	700	1.36	1.70	3500	7.18	8.97
10% ILR	1980	90	0.41	0.51	35	0.12	0.15	35	0.12	0.15	180	.65	0.81
	1990	250	1.13	1.41	100	0.34	1.42	100	0.34	0.42	500	1.81	2.26
	2000	1250	5.66	7.07	500	1.68	2.10	500	1.71	2.14	2500	9.05	11.31
	2010	1750	7.92	9.90	700	2.36	2.95	700	2.40	3.00	2500	12.68	15.85
	2020	1750	7.92	9.90	700	2.36	2.95	700	2.40	3.00	3500	12.68	15.85
35% ILR	1980	90	1.29	1.61	35	0.37	0.46	35	0.38	0.47	180	2.04	2.55
	1990	250	3.58	4.47	100	1.06	1.32	100	1.08	1.35	500	5.72	7.15
	2000	1250	17.88	22.35	500	5.32	6.65	500	5.41	6.76	2500	28.61	35.76
	2010	1750	25.04	31.30	700	7.45	9.37	700	7.57	9.46	3500	40.06	50.07
	2020	1750	25.04	31.30	700	7.45	9.37	700	7.57	9.46	3500	40.06	50.07

[a] ILR = interstitial land reduction; FLD5 = farm labor displacement (average annual income = $4,000 in 1970 dollars); FLD4 = farm labor displacement (average annual income = $5,000 in 1970 dollars); all capacity figures in MW.

TABLE 5-4. REDUCTION IN THE NUMBER OF FARM LABORERS FROM
GEOTHERMAL DEVELOPMENT—JPL SCENARIO[a]

Variables	1980	1990	2000	2010	2020
5% ILR, field, M, $5,000	0.37	1.02	5.13	7.18	7.18
5% ILR, field, M, $4,000	0.46	1.27	6.41	8.97	8.97
10% ILR, field, M, $5,000	0.65	1.81	9.05	12.68	12.68
10% ILR, field, M, $4,000	0.81	2.26	11.31	15.85	15.85
35% ILR, field, M, $5,000	2.04	5.72	28.61	40.06	40.06
35% ILR, field, M, $4,000	2.55	7.15	35.76	50.07	50.07
5% ILR, av., M, $5,000	1.35	4.77	19.18	26.84	26.84
5% ILR, av., M, $4,000	1.69	5.97	23.97	33.55	33.55
10% ILR, av., M, $5,000	2.39	6.75	33.79	47.31	47.31
10% ILR, av., M, $4,000	2.99	8.44	42.24	59.14	59.14
35% ILR, av., M, $5,000	7.55	24.00	110.42	149.61	149.61
35% ILR, av., M, $4,000	9.44	30.00	138.02	187.01	187.01

[a] ILR = interstitial land reduction; field = all field crops; av. = average crop distribution (Sheehan, 1976); M = middle Cal Tech power plant capacity Scenario (Davis, 1976); dollar values = income figures for farmworkers (1970 dollars).

lation need to be located near the energy source, owing to energy gridding in the case of residential, and portability of oil and gas supplies, for transport.

In geothermal energy, however, population located near the energy resource assumes greater importance than for fossil and nuclear energy. First, geothermal environmental effects are not as rationally diverted away from population, as in true in the case of other energy forms. Rather than prior planning, it is geological and demographic happenstance which has determined where geothermal fields are located in relation to present population. Such juxtapositions of geothermal fields and population have been discussed for Imperial County elsewhere (Pick et al., 1976; 1977) and in the two previous chapters.

Second, direct uses of hot water from geothermal resources for house heating, air conditioning, fish farming, chemical industry processing, etc., are limited to a radius around the geothermal field. This transport radius has an estimated limit of 50 miles, with present pipeline technology—a limit approached by installed space heating pipelines in Iceland. For many locations and applications, however, economics may limit transport to considerably shorter distances. As seen later in Chapter 8, in Imperial County, the dollar cost of several miles of buried pipeline may run in the millions. Thus, available population to consume residential direct uses as well as to participate in direct use industries must necessarily be local.

Historical Migration and Age Structure

Unlike many California counties, Imperial lost significant numbers of persons by out-migration from 1930 to 1970. As a consequence of the 40 year average 1.15% net out-migration rate per year, the county only increased slightly in total population from 60,903 in 1930 to 74,492 in 1970, in spite of a very high fertility level. From 1970 to 1975 this trend reversed, and the county experienced a net annual in-migration rate of 1.1% per year.

In Chapter 3, county age structures and age-specific migration were analyzed in detail for the period 1950 to 1970. To review, the following were the principal conclusions of that chapter:

1. Male shifts in age structure are difficult to analyze because of shifting border regulations on residence of Mexican citizens in Imperial County. This problem was illustrated in Figure 3-3, which superimposed the percentage age distributions for the county from 1940 to 1970. In this figure, the 1's, 2's, and 3's represent the 1950, 1960, and 1970 age structures, respectively. The dramatic drop in males aged 20–44 from 1960 to 1970 was due to the end of the Bracero Program in 1964. When the program ended, affected Imperial County residents, largely Mexican farm workers, were forced to shift their residence to Mexicali. Such anomalies make the female side of the age structure a more accurate indicator of age-specific migration.

2. As seen in Figure 3-3, the 1970 age structure had a high proportion of young age categories—44.9% under the age of 20. Such a youthful age structure is important for population projections, because large numbers of children and adolescents in 1970 will move into the prime childbearing age groups in the 1970s and 1980s adding to the potential for population increase already present from high fertility levels.

3. There was a sharp out-migration of the young, both in the 1950s and 1960's. Female ten-year out-migration rates 1950–1960 were -14.3%, -11.5%, and 10.0% for the 0–29, 30–64, and 65+ age groups, respectively. For 1960 to 1970, these respective rates were -10.3%, 0.4%, and 9.2%.

4. In accord with U.S. national trends, there has been a growing increase in older (45+) population. Such a trend is seen very clearly

on the more accurate female side of the superimposed age structures in Figure 3-3. The reasons are a combination of improved mortality and lowering of fertility levels, which leads to an eventual percentage increase in older age categories.

In summary, age structure analysis between 1950 and 1970 points to the vital importance of migration to future population. A youthful 1970 age structure and high county fertility imply large future numbers of children. The question for projection is, will these numerous young people be siphoned off, as in the past, to better economic markets elsewhere, or will they remain in the county?

Population Projections

This section presents alternative population projections, several not based on buildups in geothermal production, and others based on differing ways on projected scale-ups in geothermal electrical generation capacity. The ultimate decision as to the sequence of cause and effect between geothermal megawattage and population is left to the user of the simulations to decide. However, once such a decision is made, the simulation routine may be utilized to compute future population totals and age structures. Our hesitancy to pick one cause and effect pathway is based on lack of precedent in the United States—there is presently only one operational geothermal field, The Geysers, California, and that development differs substantially from Imperial both in the nature of its resource and in population–socioeconomic factors.

A summary of well-documented and tested population techniques gives the following major methods of population projection: mathematical methods, component methods, ratio methods, and methods taking into account economic variables (Shryock and Siegel, 1976). The present simulation is classified under the second and fourth of these. Component methods project by first determining the future levels of the growth components, usually consisting of fertility rates, mortality rates, and migration rates. These components may or may not be divided into age categories—in the present analysis, age categories are used. Simulation series (I) and (II) in the present study are direct applications of the age-specific component method. The fourth method, economic-based projection, consists of simple regression analysis employing economic independent variables, national econometric models, local econometric models derived from national models, and a limited component model based on prior projections of employment or labor force. Projection series (III–IV) of the present simulation are variations of the latter technique, in which electrical generating capacity is substituted for employment or labor force. In regard to economic projection models, "the more intensive procedures are difficult to apply since they require considerable data and involve the problem of demographic and economic independence" (Shryock and Siegel, 1976).

In the present analysis, it is acknowledged that there is a dearth of data due to lack of any operational U.S. geothermal site similar to Imperial County, and unproven demographic–economic interactions. Nevertheless, it is valuable to simulate alternative economic-based projection models in order to assess potential ranges of population affects.

Myers (1977) examined the literature on human ecology–energy interactions, and evaluated possible tools which might be used to analyze these interactions. He concluded that total population and age structure are less valuable in energy analysis than are regional population distributions. He cites the example of coal production, which in spite of advances in transportation, is still somewhat regional in end use. For reasons mentioned above, another good example of the importance of regional analysis is geothermal energy. However, it may be easier to study present regional population–energy systems than to project them into the future owing to total lack of data on regional rates of change. For this reason, the present simulation only projects for one region, Imperial County.

In a study of regional projections relative to energy consumption, the interregional migration matrix for the U.S. was projected to stability to measure present population imbalances from the stable state (Kuhn, 1977). Such imbalances were compared with present patterns of consumption and production of electrical energy and natural gas. The conclusion was that present patterns could accommodate these theoretical imbalances. This study, by its simple assumption that all projection rates will remain fixed over hundreds of years, reveals the hazards inherent in energy–population projections.

Another set of literature, discussed in Chapter 3, has focused on the intermediate network effects between energy, industrialization, and in-migration of workers. These studies reveal the complexities of energy–population interactions. In the present models, these complexities are simplified into a series of population prejections for Imperial County in which energy capacity and in-migration are directly linked. There are many ways in which such a link could take place. There will be a migration of energy construction workers into the county, since there is a limited labor pool of these specialized workers in the county. Many such workers, however, might out-migrate from the county as soon as construction is completed. As discussed in Chapter 8, however, local union workers also desire geothermal jobs.

Another possible link between energy capacity and migration involves the intermediate of increased industrialization of the county from greater availability of electrical or heat energy. In this case, the in-migrants would consist of industrial workers and their dependents. This link would have the potential to bring in much larger numbers of migrants than the limited construction workforce (See Rose, 1977 for an estimate of the construction work force per 100 MW_e of installed capacity). These and other possible energy–migration linkages are unknown for geothermal energy. The only U.S. precedent, Sonoma County, has had a limited operating history. Several of the projection models to follow assume direct proportionate or lagged proportionate relationships between energy capacity and the net county in-migration rate. The model's contribution is derived from recognition of the range of possible alternative total populations and age structures. As geothermal energy comes on-line in many localities over the next several decades, the intermediate parameters will increasingly be validated, improving the reliability and planning usefulness of the present simple and unvalidated simulation study.

Six county age-specific projection series were made. The simulation software was written in FORTRAN and incorporated the LIFE and modified PROJECT routines from Keyfitz and Flieger (1971). All six series for projection purposes assumed fixed initial fertility and mortality rates. The six series differed on migration assumptions as follows:

I. No net migration.
II. Net migration rates calculated as an average for both sexes and for the two time periods, 1950–1960 and 1960–1970.
III. The age distribution of net migration rates calculated as an average of the county distribution of rates from (II) and the U.S. national distribution of rates given in Bogue (1959). These were adjusted by scalar multiplication so that the five-year crude migration calculated on the stationary age distribution increased linearly for the first 25 years of the projection from historical 1930–1970 crude level to a level 15% of that for the five fastest growing California counties 1950 to 1970.
IV. Same as (III) except the eventual crude migration rate was 20% that for the five fastest growing California counties, 1950–1970.

V. Same as (III) except the scale-up in geothermal capacity takes place according to the Davis (1976) medium estimate of utilities' plans for scale-up in capacity, and migration effects were lagged 10 years behind scale-up.

VI. Same as (IV) except the scale-up in geothermal capacity takes place according to the Davis (1976) medium estimate of utilities' plans for scale-up in capacity, and migration effects were lagged 10 years behind scale-up.

Each projection series was run from two starting points, 1970 and 1975. For 1975, the only official datum was a total county population figure of 83,100 estimated by the State of California (1976). To obtain a 1975 age structure, it was necessary to project the 1970 population to 1975 assuming 1970 vital rates and no migration [projection series (I), that is] to yield a 1975 total projected population of 80,806. The discrepancy of this total and the state's total of 83,100 was then simply resolved by multiplying the 1975 projected age structure by the fraction 83,100/80,806. The advantages and disadvantages of the two initial years are the following:

1970—Advantages. Initial population data are from the decennial U.S. Census of Population and judged to be more accurate for 1970 than the projected data for 1975.

Disadvantages. The 1970 birth and death data used to compute vital rates for the projection is less recent.

1975—Advantages. The 1975 state figure for total population may be taken to more accurately reflect migration shifts in the 1970–1975 period, if credence is given to state intercensal migration estimates—estimates largely based on changes in automobile driver's license registrations and in school enrollments. The 1975 vital statistics are more recent.

Disadvantages. The technique of scaling the projected 1975 age structure introduces unknown errors. The 1975 state population total may be erroneous.

With six series and two starting years, twelve projection results appear in the tables. The following section analyzes pairs of projection series in more detail.

Series (I) and (II)

Series (I) is an unmodified age-specific cohort component projection. Series (II) estimates the historical five-year age-specific migration rates as follows. The four 10-year age schedules of net migration for two time periods (1950–1960), (1960–1970), and two sexes were averaged to give the historical migration schedule shown in Table 5–5. Five-year age-specific migration rates were approximated as one-half of the 10-year age-specific rates.

Series (III) and (IV)

For these two series, geothermal power plant capacity was assumed to be installed in a linear manner from 1980 to 2005. Such a linear increase in capacity has characterized the installation of 530 MW$_e$ of The Geysers' capacity between 1971 and 1978. Since The Geysers' ultimate reservoir has been estimated at four times this capacity, a 28-year installation length may be estimated for The Geysers if present trends continue. In these series, effects on in-migration were assumed to lead energy development by ten years.

The eventual net crude migration rates were assumed to be 15% for (III) and 20% for (IV) of the average five-year growth rate for California's most rapidly growing counties from 1950 to 1970. These counties were Orange (9.45% growth rate), Ventura (6.01%), Riverside (4.99%), Santa Barbara (4.98%), and San Bernardino (4.45%). Even deducting an average natural increase of about 1% per year for the five-county average, the eventual net in-migration rate of Imperial is conservatively estimated as less than a third that of the five counties. Imperial's present rural lifestyle, cultural narrowness, and noncoastal location may deter a maximal California growth pattern (Pasqualetti *et al.*, 1979).

In the projection series (III) and (IV), the crude migration rate is scaled up from the historical annual −0.01 rate to rates of +.01 and +.01 and +.0138, respectively. The computer program computes these rates by scaling the age distribution of migrants linearly until the net migration rate computed against the stationary age distribution is the assumed one. The stationary age distribution is chosen for purposes of standardization, a common demographic technique to achieve comparability (see Shryock and Siegel, 1976).

Another projection choice involves the distribution by age of migrants. Since the regions of increasing net migration rates chosen for series (III) and (IV) imply a change from Imperial's 1950–1970 net migration distribution, it was decided to average the county's 1950–1970 distribution weighted 0.3, with the U.S. age structure for intercounty mobility which appeared in Bogue (1959) weighted 0.7. All three distributions are shown in Table 5–5. This averaging assumes the likelihood in projections (III) and (IV) that Imperial will change in its future

TABLE 5-5. AGE REGIMES OF NET MIGRATION

Initial age of 5-year age group (except 85+):	0	5	10	15	20	25	30	35	40
(A) County historical net migration regime:[b,c]	123	−87	−8	−137	−254	−45	33	−3	−57
(B) U.S. mobility rates from Bogue (1959):[c]	231	163	153	197	349	312	233	184	143
(C) Average[d] of (A) and (B):[c]	108	76	145	60	95	267	266	181	86

	45	50	55	60	65	70	75	80	85+
(A) County historical net migration regime:[b,c]	−91	−80	−92	−36	32	75	75	75	75
(B) U.S. mobility rates from Bogue (1959):[c]	115	101	91	86	92	85	85	85	85
(C) Average[d] of (A) and (B):[c]	74[a]	71[a]	49[a]	50	124	160	160	160	160

[a] 50/1000 added to regime (C) rate for smoothing.
[b] Average of males and female net migration rates for 1950–1960 and 1960–1970.
[c] Expressed in mobility rate per 1000 population per year. These rates have changed only slightly in the last 30 years.
[d] Sums of absolute values of regimes (A) and (B) are in the proportions 0.33 to 0.67.

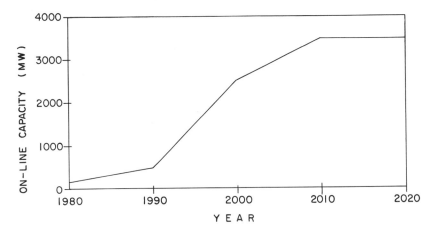

FIGURE 5-3. Projected on-line geothermal capacity, Imperial Valley, Medium growth scenario.

migratory patterns to more closely resemble an average national pattern.

Series (V) and (VI)

These series are similar in all respects to series (III) and (IV) except that the additions of geothermal plant capacity are assumed to follow Davis' (1976) medium growth scenario in capacity additions. Davis based the construction of scenarios on extensive interviews with utilities and production companies planning to build geothermal power plants in the county. Davis' medium scenario (Figure 5–3) starts gradually in 1980, and experiences a rapid buildup from 1990 to a maximum level of 3500 MW_e in 2010. Ermack's alternative scenarios are shown in Figure 5–4. The possible Imperial County locations of these plants, given different MW_e development levels, are shown in Figure 5–5.

For series (V) and (VI), change in net crude migration rate lags 10 years behind change in power plant additions. These series assume that on-line geothermal power plant capacity will have to be present prior to industrialization to attract in-migrating workers and families.

Results

Results of the total population simulation projections for Imperial County are presented in Table 5–6. The relative results between series

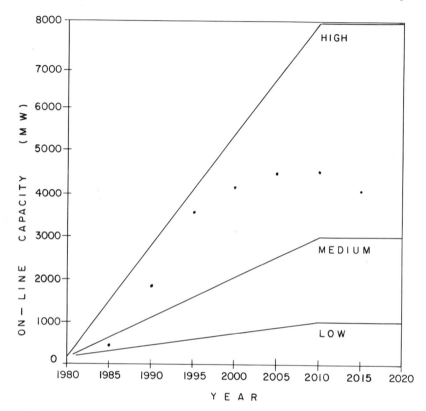

FIGURE 5-4. The projected growth rates of geothermal electric power production in Imperial Valley. Also shown (dots) is the national geothermal goal proposed by ERDA for the Imperial Valley resource. (From Ermack, 1978:205)

for the 1970 and 1975 base years are similar, except that choice of 1975 base year results in lower overall magnitudes. The reason is that the county crude birth rate dropped by 16.2% between 1970 and 1974, rising by about 5% in 1975. Thus, in the following discussion of population totals only the 1970 figures will be referred to, but the same conclusions apply in all cases to the 1975 figures. Series (I) results in a 135% population increase over 50 years. The county's potential for substantial increase, if historical migration patterns are even neutralized, is seen here. Series (II) continues the characteristic Imperial pattern since 1930. Series (III) and (IV) result in significantly larger populations than do series (V) and (VI). Such a result would indicate that if migration were to be related to power plant capacity, then the se-

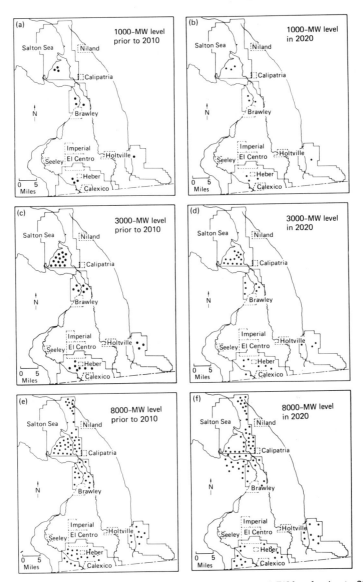

FIGURE 5-5. Geothermal power plant siting patterns: (a) 1000-MW level prior to 2010; (b) 1000-MW level in 2020; (c) 3000-MW level prior to 2010; (d) 3000-MW in 2020; (e) 8000-MW level prior to 2010; (f) 8000-MW level in 2020. (Courtesy of Donald L. Ermak, Lawrence Livermore Laboratory.)

TABLE 5-6. SUMMARY OF POPULATION PROJECTIONS FOR IMPERIAL COUNTY

Series	1970	1980	1990	2000	2010	2020
			1970 base year			
I	74,492	88,568	106,194	124,603	148,124	175,081
II	74,492	75,466	77,448	76,783	74,270	71,195
III	74,492	77,791	93,779	122,347	167,633	231,851
IV	74,492	78,359	97,789	134,740	197,387	292,570
V	74,492	80,024	85,776	92,522	107,570	136,887
VI	74,492	80,024	85,776	92,223	109,110	150,105
			1975 base year			
I	83,100	88,816	101,563	113,436	125,361	136,945
II	83,100	82,754	82,753	80,709	75,901	69,267
III	83,100	86,101	98,473	119,011	144,522	172,782
IV	83,100	86,482	101,139	126,549	159,205	196,950
V	83,100	84,294	86,216	88,197	94,993	111,368
VI	83,100	84,294	86,216	87,911	96,323	121,974

quencing of plant additions and leads or lags in translation to migration would be of vital importance for eventual total population.

It is possible that during and after a geothermal power plant scale-up, industrialization might not take place at all. So far, the scale-up in Sonoma–Lake Counties reveals very little such industrialization and resultant in-migration. In such a case, series (I) or series (II) would clearly be the proper choice.

For all series, it is also essential to note that national demographic changes may be of equal or greater importance in their effect than local factors, which might include geothermal industrialization. One recent example of such a concurrent trend is the return of population to nonmetropolitan areas since 1970 noted by Beale (1975; 1976). Imperial's shift in net migration rate in the 1970s may be a reflection of this national trend. If such a national trend were to continue and perhaps increase over the 50-year projection period, it might account for more population change than any assumed geothermal industrialization.

The projection series may also be compared on the basis of future shifts in the age structure. Table 5–7 gives the present percentage age distribution of Imperial County and the U.S. Table 5–8 presents the year 2020 age distributions of the projection series. Comparing the 1970 base year results to 1975 base year results reveals older age structures for the 1975 base year. Comparing across series, the historically determined series (II) is substantially older than any other series. It should be noted by comparison with Table 5–7 that continuance of the historical pattern implies a significant aging from the present distribution. Comparing series (III) and (IV) with series (V) and (VI) reveals a somewhat more aged age structure for the latter two. However, the difference appears too slight to be of great concern.

Goldman and Strong (1977) believe that whatever population impact comes about because of geothermal development will be within the boundaries of the seven incorporated cities of Imperial County. "The type of impact created by families indirectly related to geothermal development will be increased assessed valuation on homes, increased commercial activity and industrial expansion, and increases in both the costs and revenues for schools, cities and the county's government." They make the questionable assumption that the "unemployment rate will drop from 14.3 to 10 percent, or that 1400 jobs created directly and indirectly by the highest geothermal scenario will be filled by unemployed persons currently residing in the County." Table 5–9 shows the effects they assume.

In the operational stages, according to Goldman and Strong (1977:29–30) the fiscal impacts of the three alternatives will be as follows:

1. *Alternative 1* with five 200-MW$_e$ plants distributed fairly evenly between the four KGRAs would increase the county population by 2,000 people and the student population by 500, while increasing the county assessed valuation by $60 million. It should be noted that no commercial geothermal plants currently exist at 200-MW$_e$ capacity.

TABLE 5–7. 1970 AGE DISTRIBUTIONS OF POPULATIONS IN PERCENT

Age	Imperial County		United States	
	Male	Female	Male	Female
0–4	9.8	9.6	8.6	8.6
5–9	12.4	11.8	9.6	9.8
10–19	23.3	22.7	19.4	19.2
20–39	21.3	24.6	29.7	28.0
40–64	25.0	24.2	27.4	26.4
65+	8.0	6.8	5.3	8.0
Total number:	36,871	37,621	98,912[a]	104,300[a]

[a] In thousands.

TABLE 5-8. AGE DISTRIBUTION OF IMPERIAL COUNTY POPULATIONS IN PERCENT—PROJECTED TO YEAR 2020

Age	(I)		(II)		(III)		(IV)		(V)		(VI)	
	Male	Female	Male	Female	Male	Female	Male	Female	Male	Female	Male	Female
					1970 base year							
0-4	12.0	11.0	7.3	6.5	14.5	13.3	15.6	14.3	12.0	11.1	11.7	10.9
5-9	11.1	10.3	7.3	6.5	12.7	11.6	13.4	12.3	11.2	10.5	10.7	10.0
10-19	19.7	18.3	14.4	12.8	21.0	19.3	21.6	19.9	20.3	19.0	20.0	18.7
20-39	29.6	28.1	27.3	24.9	28.9	27.8	28.4	27.6	31.3	30.0	32.0	30.6
40-64	24.4	22.8	31.5	30.6	18.2	20.0	16.8	18.7	20.5	21.4	21.1	22.0
65+	2.9	9.2	11.9	18.4	4.4	7.6	3.8	6.9	4.7	8.0	4.6	7.9
Total number:	85,186	89,895	32,791	38,403	114,690	117,161	146,111	146,459	66,843	70,043	73,285	76,820
					1975 base year							
0-4	9.0	7.9	5.1	4.5	8.4	7.5	8.1	7.5	9.0	8.2	8.8	7.9
5-9	8.7	7.7	5.2	4.5	8.2	7.4	7.2	7.4	8.9	8.0	8.4	7.6
10-19	16.4	14.7	11.5	10.0	15.6	14.0	13.7	14.0	17.0	15.4	16.7	15.1
20-39	29.6	26.7	24.7	22.0	29.5	27.3	25.9	27.3	31.6	29.2	32.2	29.8
40-64	28.2	27.2	38.4	35.3	31.0	31.1	27.2	31.1	26.4	26.5	26.9	26.9
65+	7.9	15.5	14.7	23.4	7.3	12.7	6.4	12.7	7.0	12.6	6.9	12.6
Total number:	65,331	71,614	31,334	37,933	82,499	90,283	94,000	102,950	53,279	58,089	58,342	63,632

TABLE 5–9. EFFECT OF GEOTHERMAL DEVELOPMENT ON IMPERIAL COUNTY UNEMPLOYMENT

Alternative	Newly employed	Reduction (%)	Unemployment (%)
1: 1000 MW$_e$	175	0.5	13.8
2: 3000 MW$_e$	525	1.6	12.7
3: 8000 MW$_e$	1400	4.3	10.0

This is an arbitrary megawattage which is equivalent to form standard-sized 48–50-MW$_e$ geothermal plants. The county government would experience revenues in excess of costs of $966,000 while cities would have costs exceeding revenues by $100,000. School districts with geothermal plants would have a revenue excess of $7,000. The overall effect to county, city, and school district governments would be $3.2 million of revenues in excess of costs.

2. *Alternative 2* with 15 200-MW$_e$ plants would increase the effect experienced under alternative 1 by a factor of three (15/5= 3). However, since the growth rate in the Salton Sea KGRA is greater than the other three KGRAs, the agencies in the Northern Valley (Brawley and Calipatria, as well as the Calipatria school district) would experience growth rates and impacts exceeding those of the southern and eastern areas. The county population would increase by 6,000 people, of which 1,600 would be students. County revenues would exceed costs by $2.9 million and the cities' costs would exceed revenues by $309,000. School districts with geothermal plants would have revenues exceeding costs by $7 million if 1975–1976 tax rates were allowed to prevail, while schools with no geothermal plants would have $14,000 excess revenues over costs. The County assessed valuation would increase by $179 million. The overall effect to county, city, and school district governments would be $9.6 million of revenues in excess of costs.

3. *Alternative 3* with 40 200-MW$_e$ plants, half of which would be in the Salton Sea KGRA, would increase the effect experienced under alternative 1 by a factor of 8 (40/5= 8). The overall effect to county, city, and school district governments would be $25.5 million of revenues in excess of costs. The county's population would be boosted by 16,000 people, 4,200 of whom would be students, and the county assessed valuation would increase by an estimated $476 million. The county's revenues would be $7.7 million in excess of costs, and the cities would experience a net cost/revenue loss of $862,000, while if the

1975–1976 tax rates were allowed to prevail, the school districts would realize revenues of $18.7 million in excess of costs.

According to Goldman and Strong (1977:33–34), the three alternatives would have the same overall effect to local governments; that is, the county government and school districts would experience revenues in excess of costs, while cities would have costs in excess of revenues. As shown in Tables 5–10 and 5–11, for all three scenarios, the excess revenues in the county government would be ten times greater than the cities' cost excess. This means that for every $1.00 the county tax rate would decrease, the cities' would increase $0.10. According to Goldman and Strong, those school districts which have geothermal energy plants within the district boundaries would be required to lower their tax rates under revenue limit legislation established in SB 90 while those districts without geothermal facilities would more or less maintain the prevailing tax rate. Although their assumptions are pre-Proposition 13, their assumed school district effects seem somewhat debatable, regardless of Proposition 13 (i.e., power plants may have no effect on district tax rates).

The inequity placed on cities and some school districts without geothermal plants could be solved by establishing an intergovernment revenue transfer or by broadening local taxing jurisdictions to include a geothermal plant. In general, no school district would experience excessive losses due to geothermal energy development. Most school districts would experience major revenue gains, city governments would have service costs exceed revenues by $0.10 for every $1.00 and county government would gain in excess revenues.

Conclusions

Energy is an important variable in economic systems. If the energy type has some local aspects, such as for geothermal energy, it may affect local populations through economic effects on migration.

The present simulation study made alternative assumptions about the presence, magnitude, and sequencing of causal pathways. A limitation of the study is lack of any analogous U.S. cases on which to determine energy–population functional relationships. Nevertheless, the six projection series offer a range of demographic effects. If the assumptions of any one of these series are highly probable, the simulation may be utilized to compute prospective population totals and age distributions. In the larger multidisciplinary study, which included the present research, an input–output analysis was based on series (I) and (III) (Lofting, 1977). Imperial County geothermal plan-

TABLE 5–10. COMPARATIVE GROWTH LEVELS OF POPULATION, ASSESSED VALUATION AND COSTS AND REVENUES[a,b]

Variables	Alternative 1	Alternative 2	Alternative 3
Number of geothermal facilities (200 MW$_e$)			
Salton Sea KGRA	1	7 (×7)[c]	20 (×20)[c]
Brawley KGRA	1	3 (×3)	7 (×7)
Heber KGRA	1	3 (×3)	9 (×9)
Holtville	2	2 (×1)	4 (×2)
Total	5	15 (×3)	40 (×8)
New people	2,016	6,050 (×3)	16,130 (×8)
New students	527	1,582 (×3)	4,220 (×8)
New assessed valuation (in millions of dollars)			
For families	9.5	28.6 (×3)	76.3 (×8)
For G.T. plants	50.0	150.0 (×3)	400.0 (×8)
Total	59.5	178.6 (×3)	476.3 (×8)
New costs and revenues (in thousands of dollars)			
County government (excess revenue)	966.5	2,900.0 (×3)	7,732.0 (×8)
Cities			
Brawley (excess expenditure)	−28.0	−91.3 (×3.3)	−237.2 (×8.5)
Calexico	−29.1	−79.8 (×2.7)	−223.3 (×7.7)
Calipatria	−14.2	−54.9 (×3.9)	−144.4 (×10.2)
El Centro	−9.4	−26.7 (×2.8)	−69.3 (×7.4)
Holtville	−5.1	−14.2 (×2.8)	−36.0 (×7.1)
Imperial	−.7	−1.8 (×2.6)	−5.2 (×7.4)
Westmoreland	−13.3	−39.9 (×3.0)	−147.1 (×11.1)
Total cities	−99.8	−308.6 (×3.1)	−862.5 (×8.6)
Schools			
Brawley H.S.	197	597 (×3)	1,228 (×6)
Brawley Elementary	248	752 (×3)	1,305 (×5)
Westmoreland Elementary	−2	−6 (×3)	222 (×111)
Calexico Unified	498	580 (×1)	1,715 (×3)
Calipatria Unified	440	3,092 (×7)	9,274 (×21)
Central Unified H.S.	209	442 (×2)	1,112 (×5)
El Centro Elementary	20	54 (×3)	136 (×7)
Heber Elementary	0	256 (×25)	531 (×531)
McCabe Elementary	266	266 (×1)	799 (×3)
Holtville Unified	499	986 (×2)	2,456 (×5)
Imperial Unified	−11	−34 (×3)	−90 (×8)
Total schools	2,364	6,994 (×3)	18,688 (×8)
Counties, cities and schools grand total (excess revenue)	3,230.7	9,585.4 (×3)	25,557.5 (×8)

[a] Source: Goldman and Strong (1977:31).
[b] Minus sign (−) indicates that expenditures exceed revenues.
[c] (×) larger than alternative 1.

TABLE 5–11. COMPARATIVE GROWTH LEVELS—1975–1976 STATUS VS. THREE ALTERNATIVE SCENARIOS[a]

Description[b]	Current status 1975–1976	Current status plus alternative 1[c]	Current status plus alternative 2[c]	Current status plus alternative 3[c]
County				
Pop.	83.8	85.8 (3%)	89.8 (7%)	99.9 (19%)
A/V	290.0	349.5 (20%)	468.6 (62%)	766.3 (164%)
Cities				
Brawley				
Pop.	13.9	14.3 (3%)	15.3 (10%)	17.6 (27%)
A/V	20.8	22.9 (10%)	27.4 (32%)	38.1 (83%)
Calexico				
Pop.	13.0	13.4 (3%)	14.2 (9%)	16.2 (25%)
A/V	20.0	21.9 (9%)	25.4 (27%)	35.1 (76%)
Calipatria				
Pop.	2.1	2.2 (5%)	2.6 (24%)	3.3 (57%)
A/V	2.5	3.2 (28%)	5.0 (100%)	9.2 (268%)
El Centro				
Pop.	21.4	22.1 (3%)	23.5 (10%)	26.9 (26%)
A/V	43.2	46.6 (8%)	53.0 (23%)	68.9 (59%)
Holtville				
Pop.	4.5	4.7 (4%)	5.0 (11%)	5.7 (27%)
A/V	5.4	6.2 (15%)	7.6 (41%)	10.9 (102%)
Imperial				
Pop.	3.2	3.3 (3%)	3.5 (9%)	4.1 (28%)
A/V	5.4	5.9 (9%)	6.9 (28%)	9.5 (76%)
Westmoreland				
Pop.	1.4	1.4 (2%)	1.5 (7%)	1.8 (29%)
A/V	0.7	0.9 (28%)	1.3 (86%)	2.8 (300%)
Total cities				
Pop.	59.5	61.5 (3%)	65.6 (10%)	75.6 (27%)
A/V	98.0	107.6 (10%)	126.6 (29%)	174.3 (78%)
School Districts				
Brawley H.S.				
ADA	1585	1616 (2%)	1681 (6%)	1843 (16%)
A/V	61.12	73.71 (21%)	98.34 (61%)	140.52 (130%)
Brawley Elementary				
ADA	3449	3539 (3%)	3730 (8%)	4189 (21%)
A/V	36.55	48.61 (33%)	73.19 (100%)	103.84 (184%)
Westmoreland Elementary				
ADA	471	480 (2%)	498 (6%)	554 (18%)
A/V	11.61	11.81 (2%)	12.19 (5%)	23.72 (104%)

(cont.)

Description[b]	Current status 1975–1976	Current status plus alternative 1[c]	Current status plus alternative 2[c]	Current status plus alternative 3[c]
Calexico Unified				
ADA	5014	5125 (2%)	5334 (6%)	5884 (17%)
A/V	29.50	41.43 (40%)	44.91 (52%)	74.55 (153%)
Calipatria Unified				
ADA	1234	1262 (2%)	1336 (9%)	1502 (23%)
A/V	22.78	33.44 (47%)	95.32 (318%)	239.46 (900%)
Central Unified H.S.				
ADA	2410	2455 (2%)	2542 (5%)	2757 (14%)
A/V	79.86	93.26 (17%)	109.67 (37%)	155.52 (95%)
El Centro Elementary				
ADA	4270	4414 (3%)	4692 (10%)	5385 (26%)
A/V	48.55	51.95 (7%)	58.36 (20%)	74.21 (53%)
Heber Elementary				
ADA	732	732	732	732
A/V	3.96	3.96	13.96 (253%)	23.96 (500%)
McCabe Elementary				
ADA	271	271	271	271
A/V	12.04	22.04 (83%)	22.04 (83%)	42.04 (250%)
Holtville Unified				
ADA	1943	1985 (2%)	2062 (6%)	2253 (16%)
A/V	35.61	46.39 (30%)	57.77 (62%)	91.13 (155%)
Imperial Unified				
ADA	1562	1591 (2%)	1645 (5%)	1789 (15%)
A/V	36.00	36.52 (1%)	37.47 (4%)	40.09 (11%)
Total all districts				
ADA	22941	23470 (2%)	24586 (7%)	27159 (18%)
A/V	377.58	463.12 (23%)	623.22 (65%)	1009.04 (167%)

[a] Source: Goldman and Strong (1977:32).
[b] Pop. = population in thousands; A/V = assessed valuation in millions.
[c] Value in parentheses is the percent increase under this alternative.

ners have already incorporated results both from the population simulation and from the input–output series based on the simulation. Such simulations might be adapted and applied to other county regions with an underdeveloped geothermal resource.

Generally, principal promoters and owners of potential geothermal facilities live and work outside Imperial County (Ternes, 1978:68). Analyses of the labor pool in Imperial County also point to the conclusion that geothermal development will have relatively little impact upon the job market, especially if electrical generation is emphasized as opposed to nonelectrical uses. Since nonelectrical use is more geographically limited, such uses might overcome the unemployment that now exists in the county and add additional jobs for in-migrants. If electrical production only is carried out, only a small percentage of it will remain locally. Thus, most of the energy will be exported to the large metropolitan centers along the coast. So far, little analysis has been carried out as to who will be the beneficiary of the generated electricity. Also related to the generation of electricity is the gridding out of the county of the produced energy via transmission lines in power corridors. These power corridors so far have not been delineated. Generally, such corridors will be over 800 feet wide and stretch for miles across the county. Clearly, this also has potential population and economic impact.

Transferability of Methods and Results

The study area for this project was one county, containing less than 85,000 persons, and a geothermal resource of perhaps 15,000 MW_e maximal size. Since there are many other potentially exploitable U.S. geothermal fields—mostly in the west—one important objective of this study was to develop techniques or identify prior ones which might be applied in other localities. This section will examine each of the major methods used in Chapters 3, 4, and 5 for their transferability to other regions.

The data were mostly obtained from the 1970 U.S. Census (1973 and 1974). Since Imperial County has not changed significantly in its population characteristics since the 1930s, 1970 data is considered relevant in 1977. However, in geothermal areas of rapid population growth, the population change might well invalidate use of 1970 data seven years later. In such a case, the data for a study like the present one would be unavailable—at least until the 1980 Census results are released in the period 1981–1984. In off-census years these faster-growth areas would have to rely on special federal, state, or local interim censuses or surveys, which would likely have county or city areas as the smallest units and which would probably be much less accurate than the decennial federal census. Even for historically static counties, such as Imperial, it may be advantageous to use more recent data than the last federal census. However, the plus of greater recency of such data should be weighed carefully against the minus of likely greater inaccuracies in sampling and data collection.

In the 1970 Census, urban areas—those contained within the census-defined regions, urbanized areas and standard metropolitan statistical areas (SMSA)—have a much larger and more detailed body of published data on population characteristics. For a geothermal study in an urbanized area or SMSA, greater number and detail of characteristics could be investigated. The fifth count ED data coverage would remain the same (See the Introduction to Chapter 4 for explanation of ED data). Another problem in urban areas would be choice of population affected by geothermal development. If, for example, a large geothermal field were located 50 miles from the Denver city limits, the choice of population susceptible to, say, nonelectrical or noise effects would be considerably more difficult to delineate than in Imperial County.

Another important data problem to the present study is that of seasonality in certain socioeconomic characteristics and seasonality in dates for census collection. An example from Imperial County is measurement of the number of farm laborers. It makes a significant difference whether such data are collected in the maximal harvest times of May and January, or in off-harvest times such as February to mid-April. Another example of such seasonality is measurement of vacancy rates in a county with both winter tourists and considerable summer travel away of residents, due to the extremely hot weather. Although Imperial County is far more seasonal than most, researchers anticipating similar studies in other areas should carefully check variables for such seasonal tendencies in order to improve and strengthen their analysis.

In the present study, the key socioeconomic variables turned out to be ethnicity, age structure, fertility, migration, household density, and a few others. Prior to research, there was little indication which variables would be most important. Likely, other geothermal areas will have a partially different list of key variables; thus, it may be necessary to begin with a sizable list of variables in order gradually to identify key ones.

Assuming that currently applicable ED data are available, the statistical methods of regional computer graphics, discriminant analysis, correlation, and regression analysis are transferable to studies of other areas. In preparing computer maps of rural areas, ED boundary maps, drawn by local governments, may contain certain errors or inadequacies. For mapping, there are several line-printer software routines available, such as AUTOMAP II and SYMAP. Another possibility is to digitize the map boundaries, and use a plotter and associated mapping software for a finer plot. Choice would depend on the researcher's computer costing and core size, as well as the desire for fineness of detail. For computer plotting of population by age category, routines such as PYRAMID (Pick, 1974) and PLATO (Handler, 1972) are available.

For regression analysis, the common statistical problem of choice of independent and dependent variables presents itself—choices dependent on understanding of the key socioeconomic processes potentially affecting geothermal development. Two other well-documented problems in design of regressions and discriminant analyses are errors due to high intercorrelations of independent variables (nonorthogonality) and choice of stepwise versus nonstepwise techniques. Detailed information on these sources of error are contained in Lachenbruch (1975) and in standard statistical textbooks.

The estimation of farm labor displacement due to geothermal land use would appear applicable elsewhere only in agricultural areas. Few industries other than agriculture have work forces with a direct relationship to land area. Even in agricultural areas, the analysis is complicated for projection purposes by future crop and farm worker productivity changes. Also, geothermal development may cause indirect land losses, such as those influenced by geothermal-related industrialization. Another problem is the large number of geothermal unknowns, such as well spacing and subsidence. It is due to these unknowns that *ranges* of interstitial land losses must be estimated.

The population projection methodology is based on standard techniques (Keyfitz *et al.*, 1971), and appears transferable to other areas. In Imperial County, as elsewhere, the major projection unknowns are fertility and migration values. Since fertility is so complicated and difficult to predict, demographers often end up assuming the latest fertility values for the entire length of the projection. Cyclical U.S. fertility shifts in the first half of the 20th Century have been studied by Easterlin (1961), and cyclical projection assumptions for fertility might be a wiser choice for another county, although fixed fertility was assumed for Imperial.

Projection assumptions on future migration generally require thorough understanding of historical volume and age structure of migrants. Three types of migration must be distinguished: in-migration (persons entering a region over a time period), out-migration (persons leaving a region over a time period), and net migration (in-migration minus out-migration over a time period). One difficulty in standard projections is that net migration values are used, although in-migration and out-migration are much more explainable as individual socioeconomic processes. Values for net migration, which is a difference between in- and out-, may be less meaningful. In a scenario for a geothermal region in which the economy is being fundamentally altered, it is very difficult to estimate the future age distribution for net migration, as even the initial values may not make sense. The answer would appear to lie in closely studying in- and outmigration for the region, with the restriction that much more census data are available for in-migration than out-migration.

The transferability of specific Imperial County results to other regions is problematic since there are many unique features to the county, such as proximity to Mexico, young age structure, high Spanish-American population, and others. Therefore, if possible, it would appear better to repeat our analyses rather than transferring specific results directly.

6

Public Opinion about Geothermal Development

INTRODUCTION

A recent review of public involvement in planning public works projects carried out by the Comptroller General's Office of the U.S. Government suggested that public participation in the government's decision-making process has long been recognized as a necessity for planning and developing public works projects (1974:i). This report is relevant to Imperial County's geothermal development, which, although stemming from private sector projects, has an unusually strong governmental involvement. The Comptroller General's Office feels that traditional public hearings do not allow the public to express its opinions and do not provide a good forum for evaluating and discussing issues and policy alternatives. Thus, there is a belief that the public hearing process is insufficient and that an effective public involvement effort should make sure of the following:

(1) The public has an opportunity to be heard earlier, before major project decisions are made.
(2) Adequate notice of opportunities for involvement is provided to interested or potentially affected parties.
(3) Frequent forums are held throughout all stages of project development.

As a result of various surveys carried out by the Comptroller General's Office, they suggested that all projects affecting the public ensure that citizens potentially affected by, or interested in, a project be identified and directly notified of their involvement opportunities.

Some of the departments that the Comptroller General's Office surveyed believed that such requirements possibly may be unrealistic because of the following: (1) difficulties in precisely identifying large and complex project areas and the persons affected, (2) cost of large mailings, (3) possibility of legal action by persons inadvertently omitted from the mailing list. In the Comptroller General's Office opinion, these were not valid reasons for not using direct mail to notify the public.

The Comptroller General's Office assumed that the following would improve the credibility of public works projects: (1) the planning process should be open to public scrutiny, (2) the public's views should be objectively considered, (3) all known relevant data and information should be shared with the public, (4) citizens should feel that their interests are being objectively and fairly represented throughout all stages of project development (1974:5). While geothermal development cannot be categorized as "public works," certainly these views apply to such developments.

One mechanism to ensure public participation is *public opinion surveys*. This chapter reports the results of a public opinion survey primarily concerned with geothermal development conducted in Imperial County, California, in 1976. The two original stated primary tasks of this research component were as follows:

(1) To investigate community attitudes in order to anticipate any problems that geothermal development might bring about in Imperial County.
(2) To determine if there are differences of opinion in regards to geothermal development among different population categories of residents in Imperial County—e.g., citizens as opposed to leaders and landowners vs. nonlandowners.

Subsequently, major research emphasis was placed on investigating the community attitude component and using the information obtained from the complete cross section of residents of Imperial County to determine if there were differences of opinion between citizens and those in leadership positions. The survey also had as its goal ascertaining agreement and disagreement among crucial social issues involved in geothermal development, especially those that have the potential of engendering conflict between energy-related and agricultural interests and among various population segments, e.g., race/ethnicity and social status—leader vs. nonleader, landowners vs. nonlandowners, etc.

This chapter is divided into four major sections. The first section describes the research methodology, while the second section presents an analysis of public opinion survey data obtained from a scientific cross-section sample of all known Imperial County households. In the sample survey field, the term cross section refers to the inclusion in the sample of all population sections in numbers proportionate to their presence in the overall population. The second section examines data related to leadership positions and public opinion about geothermal development, as well as presenting some general speculation about subsequent relationships among citizens, leaders, and, perhaps, developing leadership networks. The final section presents some tentative conclusions, offers some policy alternatives and recommendations, discusses the extent of the transferability and utilization of this research in other geothermal areas, and briefly outlines some future research that might be undertaken in Imperial County as geothermal development progresses.

METHODOLOGY

Sample Design

For the purpose of drawing a sample of respondents in Imperial County, California, to answer a self-administered questionnaire on issues concerning geothermal development, it was necessary to construct a population pool which would include as nearly as possible every household in Imperial County. The preferred household interview technique was judged too costly for this study.

The first step in the procedure was to obtain the most recent criss-cross telephone index; therefore, a 1975 issue of *Polk's Imperial Valley Directory* was secured. Error is introduced by the necessity for telephone numbers. This error may be larger in Imperial County than statewide, as there is likely a larger proportion of transient population than for the state, especially within the County's large farm laborer sector. However, no better criss-cross directories were available for the study. The section of this directory listing the streets of Brawley, Calexico, El Centro, Holtville, Imperial, Seeley, and Westmoreland, was rearranged providing a listing of streets in alphabetical order, with each street name followed by addresses, telephone numbers, and name of primary residents. Unoccupied dwellings were denoted as vacant; businesses were indicated by firm names, and public buildings were so designated. By this method, it was possible to identify all residential addresses, and the names of presumable heads-of-households. All streets and addresses were separated into the seven above-mentioned towns and communities.

There was concern that all households in Imperial County be included in the pool, including all rural residents, so the next phase involved direct communication with postmasters in each community. An inquiry was made to determine if households outside city limits had box numbers instead of street addresses and, particularly, if the central offices serviced any star or rural routes.

The postmasters of Calexico, El Centro, Holtville, and Imperial explained that all households, including rural, in their service areas have street addresses. Mail so addressed is delivered to the home by carrier. Brawley's postmaster indicated that his office provides this same type of service, but in addition, his office serves the town of Glamis, located about 30 miles easy of Brawley. Unsure about actual number of households, he estimated slightly over 30 permanent families in that area. Mail is delivered to Glamis through his office.

The postmaster of Seeley, on the other hand, said that all households have box numbers and receive mail only at the central office. Box numbers are random; she could not provide a list of the approximately 250 occupied boxes at her office. Since mail addressed to a street address would reach the resident through assignation to the post office box, it was decided to keep the home address list obtained from the criss-cross directory.

Westmoreland proved to be a different situation. The Westmoreland postmaster explained that all households have box numbers in numerical order, with the restriction that no box numbers end in either number nine or zero. A list of all numbers from 1 through 788 was drawn up and substituted for street addresses from the directory. The Westmoreland post office also serves a star route consisting of 11 households. The postmaster provided a list of these nonconsecutive box numbers, which were added to the pool.

The next step in the process was to contact postmasters of towns not listed in the index. Ocotillo and Heber both have a consecutive numbering system of post office boxes for all households in their areas, all of which were currently occupied. Neither office provides home deliveries, nor do they service rural or star routes. As a result, 299 box numbers for Ocotillo and 606 for Heber were added to the population pool.

Calipatria, Niland, and Winterhaven have no home deliveries for "urban " householders, either, but each post office provides home delivery to rural route patrons. Postmasters of each of these towns explained that their numbering system of boxes is not consecutive, but numbers were "scattered," therefore, a beginning and ending number would not be useful. The total number of occupied central post office boxes was obtained from each postmaster, as well as the total number of rural patrons serviced by each office.

Winterhaven's postmaster suggested that the small town of Bard not be overlooked, which was fortunate. Up to that point, no other postmaster had mentioned the town, although several had read the list of Imperial County places, in an effort to determine if all post offices were included; none had noticed its absence. We were assured there were no others. Bard is a town of 142 postal patrons, and the numbering system is not in rotation. This completed the construction of the household population pool for the county.

The next important step was to determine the size of the sample to be drawn. A sample of about 1200 households was adequate for our purposes. [See Jessen (1978) for a full explanation of survey techniques.] Based on 1970 United States census data, the total county population of approximately 77,000 residents lived in an estimated 19,200 households, assuming an average of four persons per household. In order to meet requirements for obtaining a sample size of approximately 1200 respondents, one out of every 16 households in Imperial County was sent a questionnaire by mail. The response rate of about 25% is discussed below.

The beginning number of three was randomly drawn, which meant that for the cities of Brawley, Calexico, El Centro, Holtville, and Imperial, the count began at the third residential address on our lists, and subsequently every 16th address was selected. The Heber, Ocotillo, and Westmoreland samples were selected in the same manner from the lists of urban and rural post office numbers which were drawn up earlier. Only residential addresses were included in the count; if the 16th number happened to be a business or public building, the next first residence on the list was selected. If the house number was listed as vacant, the assumption was made that the residence was vacant only temporarily at the time of directory enumeration and probably was currently occupied. For these addresses, the envelope was addressed to "Resident."

Respondents in all remaining cities and towns receiving service exclusively by post office box or rural route number, including Glamis, were mailed a parcel containing envelopes addressed "Box Holder," the city's name, and zip code number. The number of envelopes provided was determined by the total number of patrons on the basis of one for every 16th box holder. Each parcel was accompanied by an attached letter to the postmaster of each city and town requesting cancellation and distribution according to instructions. A form was enclosed on which the postmaster indicated the box numbers in which each envelope was deposited. A self-addressed, stamped envelope was provided for return of the completed form. This information was imperative for any followups.

We are confident that virtually all households, including rural, in Imperial County were included in the pool from which the sample was drawn. Every possible effort was extended to ensure that no household was overlooked, and that a good response from this random sample would provide a representative, reliable, and/or unbiased estimate of the attitudes and beliefs of Imperial County residents about geothermal development.

Questionnaire Design and Review

The questionnaire evolved out of a number of conferences with Imperial County geothermal project scientists and by perusing previous environmental research literature. The questionnaire used by the Lawrence Berkeley Laboratory to study public opinion about geothermal development in Cobb Valley and Lake County, California, also influenced the Imperial County questionnaire (Vollintine and Weres, March, 1976; June, 1976). The Imperial County questionnaire covered virtually all of the topical areas in those questionnaires as well as adding others.

Scientists on the Imperial County Project had the opportunity to review the questionnaire several times; critiques were received from Imperial County government staff and NSF staff, and it was reviewed by several officers of the Sierra Club. The extensive review of the questionnaire resulted in some changes, especially in the "attitudinal" statements—the final version of which are shown as statements numbered 127–149 in the Appendix. Also, the final English language version of the questionnaire and letter of introduction used in the research are shown in the Appendix. Although not shown, a Spanish language version of this material exists and was used also.

Mail-Back Procedures

Each sample household was sent a questionnaire and an introductory letter. If a questionnaire was not returned in an approximate two-month period, another questionnaire was sent with a request for a response. Despite our every effort in utilizing the *City Directory* for Imperial County and obtaining post office box numbers from the various cities and towns in Imperial County from the postmasters, a substantial number of questionnaires were returned with varying reasons for lack of delivery. Many of these places in which the questionnaires were not delivered were later ascertained to be vacant while the reason others were returned remains problematic.

Response Rates

Over 300 questionnaires were returned from Imperial County residents, with an additional 50 or so to be allocated to the completed, vacant, and other categories. However, the data analysis included in this report uses a base N of 269.

Cautions

It is appropriate to view returned questionnaires as reflecting only the opinions of persons who took the time to return them. However, the extent of agreement on some major issues involved in geothermal development is such that it is highly likely that the results reported in this chapter accurately reflect the general opinion of Imperial County residents (however, see the section on public opinion in the Heber area).

PUBLIC OPINION ABOUT GEOTHERMAL DEVELOPMENT IN IMPERIAL COUNTY: 1976

As shown in Table 6–1, only a very small proportion of the population (19.0%) feel that they have a very good understanding of geothermal development; 61% of the population feel that they only have an average or less understanding. Yet, Table 6–2 reports that most of the population in Imperial County currently are in favor of geothermal development with almost 90% reporting themselves as being in favor or strongly in favor.† This is in contrast to the 65% of Cobb Valley residents and the 75% of other Lake County, California, residents who supported geothermal development (Vollintine and Weres, June 1976:13). More people reported themselves to be in favor than they felt others were. They also felt that almost one out of four people did not care one way or another about development. Most peo-

---

†This is not an unusual occurrence; more than 88% of responses were favorable to building the Kaparowitz Generating Station in nearby Arizona and Utah communities (Little and Lovejoy, 1977). This positive stance was based on the belief that the project would bring money and jobs to local residents; a similar belief exists in Imperial County.

TABLE 6-1. UNDERSTANDING OF GEOTHERMAL DEVELOPMENT

Understanding	Frequency	Frequency (%)
Very good	51	19.0
Fair	88	32.7
Average	76	28.3
Slight	40	14.9
None	10	3.7
No response	4	1.5
Total:	269	100.1[a]

[a] Rounding error.

ple believe that civic leaders in Imperial County are in favor of geothermal development.

As shown in Table 6-3, most people agreed or strongly agreed that geothermal development should be strictly regulated. This is similar to Cobb Valley, where 84% of the population was in favor of strict regulations. Only *one person* felt that development should be completely prohibited, while less than 10% felt that geothermal development could be completely unregulated, and four out of five wanted close regulation. To about three out of every four people in Imperial County who responded to the survey, preserving the environment was an important issue. Only about one out of every five people saw any environmental or social problems arising out of geothermal development. For those people who thought there would be problems, most of them felt that there would be environmental problems as opposed

TABLE 6-2. IN FAVOR OR OPPOSED TO GEOTHERMAL DEVELOPMENT IN IMPERIAL VALLEY?

In favor or opposed	Frequency	Frequency (%)
Strongly in favor	110	40.9
In favor	131	48.7
Don't care	15	5.6
Opposed	6	2.2
Strongly opposed	1	0.4
No response	6	2.2
Total:	269	100.0

TABLE 6-3. REGULATION OF GEOTHERMAL
DEVELOPMENT

Regulated or not	Frequency	Frequency (%)
Strongly agree	69	25.7
Agree	134	49.8
Uncertain	39	14.5
Disagree	16	5.9
Strongly disagree	4	1.5
No response	7	2.6
Total:	269	100.0

to social problems such as population increase, need for more services, the use of tax money, and so forth.

According to our survey questionnaire responses, there is relatively little opposition to geothermal development in Imperial County. However, about 15% of the respondents believe that there is some opposition in Imperial County to geothermal development. The opposition to geothermal development, according to these responses, apparently is unorganized and no discernible pattern exists in regard to social characteristics of individuals or organizations that might be opposed.

While about one out of every four believed that there had been meetings held to explain and discuss geothermal development in Imperial County, only about 10% of the Imperial County respondents reported that they had attended a meeting on potential geothermal development. Of those who attended a meeting on geothermal development, the vast majority of them learned about the meeting through the newspaper.

About one out of every three respondents reported that they had visited a geothermal testing site. The majority of respondents—well over 80%, reported that no one had ever asked them how they felt about geothermal development in Imperial County. Table 6-4 shows that only 5.9% of the population feel that citizens of Imperial Valley have been adequately informed about geothermal development. Thus, there clearly is a need for an educational program about geothermal energy and development in Imperial County.

An open-ended question asked "What do you expect from geothermal development?" Generally, responses clustered into several different areas. Among the most important ones were the following: (1)

TABLE 6-4. HAVE CITIZENS OF IMPERIAL
VALLEY BEEN ADEQUATELY INFORMED ABOUT
GEOTHERMAL DEVELOPMENT?

	Frequency	Frequency (%)
Yes	16	5.9
No	179	66.5
Unsure	68	25.3
No response	6	2.2
Total:	269	99.9[a]

[a] Rounding error.

about one out of five persons believed that geothermal development would result in a cheaper source of energy; (2) a little under 20% of the population felt that it would bring about more energy availability to Imperial County; (3) and about 15% mentioned a variety of miscellaneous but positive benefits. Other less often reported expectations from geothermal development in Imperial County were more industry and jobs, some minor environmental changes, and so forth. Only approximately 5% of the population had negative expectations. Another question was asked about what effect geothermal development would have upon property values. Well over half of the population thought that property values would somewhat or greatly increase.

A variety of statements were given, with which respondents were asked to express agreement or disagreement, in regards to geothermal development and its potential impact upon Imperial County; the responses were shown in Table 6-5 (see the Appendix for the entire

TABLE 6-5. PUBLIC OPINION ABOUT GEOTHERMAL DEVELOPMENT IN
IMPERIAL COUNTY, CALIFORNIA

Statements[a]	Strongly agree/ agree (%)
127. Geothermal development will bring new tax revenues to Imperial County. (+)	66.2
128. Noise from geothermal development can be bothersome. (−)	13.0
129. Economic benefits from geothermal development are more important than environmental costs.	37.2
130. Because it will attract new residents, I'm against geothermal development. (−)	3.7

(cont.)

Statements[a]	Strongly agree/agree (%)
131. The construction of geothermal power plants, transmission lines, pipelines, and roads which result from geothermal development will create eyesores. (−)	13.8
132. Because it will attract new businesses and help Imperial Valley grow, I'm in favor of geothermal development. (+)	75.1
133. Most geothermal electricity produced in Imperial County should be used in Imperial County.	52.8
134. A fuel shortage will develop in the United States unless geothermal and other sources of energy are developed. (+)	73.6
135. Geothermal energy will provide cheap electricity for Imperial Valley.	42.4
136. I like Imperial Valley the way it is, and don't want it to change.	7.4
137. New developments like geothermal are not welcome in Imperial County. (−)	5.9
138. Most geothermal electricity produced in Imperial County will be used in Imperial County.	22.7
139. Imperial County can broaden its economic emphasis to more than agriculture through geothermal development. (+)	73.6
140. Geothermal companies should have the main responsibility to plan and conduct steam exploration and production.	53.9
141. Geothermal development may cause unusual odor problems. (−)	8.9
142. Geothermal development will increase demands on city and county government and thus increase taxes. (−)	16.0
143. Geothermal development will increase jobs in Imperial County. (+)	81.0
144. Local government officials have primary responsibility to plan geothermal exploration and production.	33.5
145. Geothermal development will take water away from agriculture. (−)	4.8
146. Geothermal resources in Imperial Valley should be used for purposes other than electricity, such as by industry or for chemicals.	39.4
147. Geothermal development will result in fewer Mexican National agricultural workers crossing daily into Imperial Valley.	7.1
148. The Imperial County policy that new industries, like geothermal, should be able to live with agriculture is a good one. (+)	82.2
149. Geothermal development will cause border regulations to change making it easier for Mexican National workers to cross into the United States.	5.9

[a] (+) Clearly pro-geothermal-development statements; (−) clearly anti-geothermal-development statements.

questionnaire). Generally, responses to each of these statements are in harmony with the earlier reported responses of favorability of the general population in Imperial County toward geothermal development. Only a few responses indicated that people were not in positive agreement about geothermal development. Those examined in detail below are only those few statements with some variation in responses.

First, there was ambivalence to the statement about economic benefits of geothermal development being more important than environmental costs (q. 129). Approximately four out of every ten people agreed with that statement. Yet, also almost one out of three disagreed with the statement, indicating a split opinion in this regard.

Similarly, varied responses were obtained to the following statement: "Geothermal development may cause unusual odor problems" (q. 141). About six out of every ten respondents were uncertain, illustrating a lack of knowledge about geothermal development and potential pollution problems. For those who responded other than uncertain, over 30% of them disagreed with this statement.

Again, with the exception of these very few statements which indicated uncertainty or ambivalence, all the rest of the responses to these statements about geothermal development show that the population in Imperial County is substantially in favor of geothermal development.

Understanding of Geothermal Development

Background Characteristics

There were no reported differences in understanding of geothermal development (q. 107) and where the person lived before Imperial County, how long they lived in Imperial County, how long they have lived in their current house, and whether they planned to remain in Imperial County. Similarly, there were no differences in understanding reported by rural or town land ownership, type of residence, tenure status, and the number of people in the household, or how many months the person had worked during the previous year.

The major differences noted in reported understanding of geothermal development was by multiple indicators of social status, and the results were consistent. For example, people with higher level educations were more likely to report a good understanding of geothermal development than those with an average or lower level of education. Residents who reported having an average or better understanding of geothermal development were more likely to have a middle or higher level of occupation while those who reported a slight or no understanding were more likely to have lower level educations. Similarly, those with higher level incomes were more likely than those with middle or lower level incomes to report a good or better understanding of geothermal development.

There was significant variation in reported understanding of geothermal development by race/ethnicity; those most likely to report a good or better understanding were Asian/Americans, Indians, and people who responded that their race/ethnicity was "other." A greater percentage of anglos than chicanos reported having a good or better understanding, 53.4% and 40%, respectively. A greater percentage of residents who were citizens of Mexico reported a slight or no understanding than chicanos who were United States citizens. Only two black persons responded to the survey and one reported a slight understanding while the other reported no understanding of geothermal development. Although only a few percent of the county's population is black, the black response rate is very low. Such a low response rate would undoubtedly have been improved by the household interview technique. The problem of nonresponse to our survey is also apparent in the special survey of Heber residents discussed in a later section. However, our primary goal was to access overall county-wide opinion. Since strong agreement exists on major issues of geothermal development, our results are highly likely to reflect overall county opinion.

Interestingly, there was a decline in reported understanding of geothermal development by the time it took a respondent to return the questionnaire, with those returning earliest having the greatest understanding with a constant decline as questionnaires were returned.

Attitudes

Slightly over half of the population reported a fair or better understanding of geothermal development. In examining the understanding of geothermal development with being in favor or in opposition, the small proportion of the population who did not care or were opposed to geothermal development were more likely to have only an average or less understanding of geothermal development. Thus, level of understanding of geothermal development in Imperial County is statistically related to being in favor or opposed to it. A similar statistical relationship holds for whether or not one believes that most people in Imperial County are in favor or are opposed to geothermal development (q. 109), and whether or not most civic leaders are in favor or opposed (q. 110).

At least four out of five persons who had a good or fair understanding of geothermal development agree that it should be strictly regulated; whereas about three out of five who have a slight or no understanding favor strict regulation. Apparently, knowledge about geothermal development increases one's belief in the necessity for strict regulation. The greater one's reported knowledge about geothermal development the less likely they are to report that property values will increase. Generally those less knowledgeable are least likely to report expectations for geothermal development or to have neutral or miscellaneous expectations.

Table 6–6 reports the expected benefits of geothermal development by understanding. Residents expect cheaper electricity. People with an average understanding of geothermal development report that they expect cheaper electricity and a mixed variety of other positive benefits. Those with a very good or fair understanding are most likely

TABLE 6–6. EXPECTATIONS OF GEOTHERMAL DEVELOPMENT IN IMPERIAL COUNTY BY UNDERSTANDING

Expected benefits	Understanding		
	Very good/fair (%)	Average (%)	Slight/none (%)
Cheaper electricity	43.9	36.8	36.0
More industry/jobs	9.4	7.9	6.0
Industry using other sources of energy	2.9	6.6	0.0
Environmental changes	6.5	3.9	4.0
Miscellaneous positive benefits	12.2	18.4	4.0
Miscellaneous neutral comments	5.0	5.3	8.0
Miscellaneous negative comments	7.2	5.3	0.0
Miscellaneous	2.2	3.9	8.0
No response	10.8	11.8	34.0
Totals:	100.1[a]	99.9[a]	100.0

[a]Rounding error.

TABLE 6–7. ENVIRONMENTAL OR SOCIAL PROBLEMS

Foreseen environmental or social problems arising out of geothermal development	Understanding of geothermal development		
	Very good/fair (%)	Average (%)	Slight/none (%)
Yes/unsure	43.9	48.7	64.6
No	56.1	51.3	35.4
Totals:	100.0	100.0	100.0

to report an expectation of cheaper electricity, other positive miscellaneous benefits, and more industry and jobs; overall they were most likely to respond to this open-ended question (q. 114). Generally, there were a very few negative expectations recorded, although those with a very good/fair or average understanding were more likely than those who had a slight/no understanding to report a negative expectation.

As shown in Table 6–7, level of understanding also is statistically related to whether or not a person foresees environmental or social problems arising out of geothermal development (q. 116); an increasing likelihood of problems is associated with a reported average or slight/none understanding of geothermal development.

No statistically discernible differences exist by understanding and awareness of opposition to geothermal development (q. 118). Those with a better understanding of geothermal development are more likely than others to report knowing about meetings held to discuss it (q. 121), although they were openly slightly more likely to have attended such a meeting (not a statistically significant difference). There were no differences in how one found out about such a meeting nor in level of understanding or in whether or not the respondent felt that citizens of Imperial County had been adequately informed about geothermal development. Almost seven out of ten residents, regardless of their understanding, reported that citizens had not been adequately informed about geothermal development. Those who report an adequate or better understanding were more likely to have visited a geothermal development site than those who had an average or slight/no understanding (q. 126).

Insofar as attitudinal dimensions are concerned and level of geothermal understanding, there were no differences reported for eleven of the questions. However, on two questions there were statistically significant differences by level of understanding of geothermal development.† On questions 144 and 146 there apparently is substantial confusion since the differences in understanding of geothermal development have mixed but statistically significant responses. Thus, agreement or disagreement with the statement that "Local government officials have primary responsibility to plan geothermal exploration and production" (q. 144) shows that those who had an average or better understanding were more likely to agree with this statement, while the most likely response for those who had a good or better understanding of geothermal development. Yet those who had an average level were most likely to agree with it and those who had slight or no understanding were most likely to be uncertain. Thus, there is substantial confusion around who should have primary responsibility to plan geothermal exploration and production insofar as it is related to current understanding of geothermal development.

Similarly, there is confusion revolving around the statement that "Geothermal resources in Imperial Valley should be used for purposes other than electricity, such as by industry or for chemicals" (q. 146). The most likely response of those who had a fair or better understanding of geothermal development was to agree with this statement. Residents with an average understanding were most likely to agree; people who had a slight or no understanding of geothermal development were more likely to respond as being uncertain. Thus, there is some confusion among Imperial County residents regarding the uses of geothermal power for purposes other than electricity.

All of the remaining questions were statistically significant in regards to level of understanding of geothermal development. Without exception, the greater the level of understanding the greater the likelihood of agreement with positive statements about geothermal development in Imperial County. For example, residents who had an average or better understanding of geothermal development substantially disagreed with the statement that "Geothermal development will take water away from agriculture" (q. 145), whereas of those who had slight or no understanding, about half were uncertain about 40% disagreed with this statement. Thus, no matter what the level of understanding of geothermal development, at least half agreed with this statement; however, the greater the level of understanding, the more substantial agreement there was.

Agreement or disagreement that "Noise from geothermal devel-

†There were no significant differences on all of the other statements. In subsequent discussions, only those questions and statements with a statistically significant difference will be discussed.

opment can be bothersome" (q. 128) had somewhat of a mixed response. Those who had the highest level of understanding of geothermal development were most likely to disagree with this statement, whereas those who had an average or less understanding were uncertain as to noise effects from geothermal development. Again, understanding of geothermal development is associated with a positive attitude toward it. Similarly, level of understanding leads to a greater likelihood of agreement that "Economic benefits from geothermal development are more important than environmental costs" (q. 129).

While virtually everyone agrees that "Because it will attract new business and help Imperial Valley grow, I'm in favor of geothermal development" (q. 132), those who have an average understanding are more likely to agree than those with a slight or no understanding or those with a fair or very good understanding. Again, however, at least three out of every four persons agree, so the variation is only in the extent of agreement. Similarly, while almost three out of every four persons agree that "A fuel shortage will develop in the U.S. unless geothermal and other sources of energy are developed" (q. 134), agreement increases with a greater understanding of geothermal development.

As with the two previous statements, a substantial majority of the population in Imperial County agrees that "Imperial County can broaden its economic emphasis to more than agriculture through geothermal development" (q. 139). However, the greater the reported understanding of geothermal development, the more likely the person is to agree with this statement. For example, three out of every five people who have a slight or no knowledge of geothermal development agree, whereas four out of every five persons who have an average or very good understanding of geothermal development agree with this statement.

Those with average or better understanding are most likely to disagree with statements relating to making it easier for Mexican National agricultural workers to cross into the United States, whereas those with slight or no understanding of geothermal development are those least likely to disagree with these statements (q. 147 and q. 149).

Finally, there is substantial agreement that "The Imperial County policy that new industries, like geothermal, should be able to live with agriculture is a good one" (q. 148). The only variation that exists is in the extent of agreement, with at least 70% of those who have slight or no understanding of geothermal development agreeing, whereas well over 80% and approximately 90% with average to fair or good understanding agree.

Overall, for about half the statements there is no difference in understanding of geothermal development and potential benefits or impact upon Imperial County. For the other statements, there are differences by level of understanding, and invariably a greater understanding of geothermal development is associated with a person being more in favor of geothermal development in Imperial County. Responses to two statements suggest confusion about potential affects of geothermal development upon Imperial County. In conclusion, the strong impression is that knowledge about geothermal development results in a positive attitude toward geothermal development in Imperial County.

Adequacy of Information about Geothermal Development

Background Characteristics

Of all the demographic and social background characteristics, only time in responding to the questionnaire is statistically related to whether or not a person feels that citizens of Imperial County have been adequately informed about geothermal development: first responders were more likely to report no, while subsequent respondents were more likely to report themselves as being unsure about the adequacy of information about geothermal development.

Attitudes

As indicated earlier, Imperial County residents who feel adequately informed about geothermal development (q. 124) are likely to report also having a better than average understanding about it (q. 107). While most of the population of Imperial county is in favor of geothermal development as shown in Table 6–8, residents who feel that they have been adequately informed about it statistically have a greater likelihood of being in favor or opposed, while those who do not have adequate information are most likely to report themselves as not caring about geothermal development. Similarly, Table 6–9 shows that residents who report citizens of Imperial County have been adequately informed about geothermal development are about equally divided about strict regulation of geothermal development, whereas those unsure or reporting that citizens are not adequately informed are substantially in favor of strict regulation (a similar statistical result held for responses to q. 112).

TABLE 6-8. ADEQUATE INFORMATION ABOUT
GEOTHERMAL DEVELOPMENT AND VIEWS ON
STRICT REGULATION

	Adequate information (Q. 124)		
Agree or disagree on strict regulation	Yes (%)	No (%)	Unsure (%)
Agree	43.8	79.4	79.4
Uncertain	12.5	16.6	11.8
Disagree	43.8	4.0	8.8
Totals:	100.1[a]	100.0	100.0
	(N = 16)	(N = 175)	(N = 68)

[a] Rounding error.

No statistically discernible differences were found in adequacy of information and expectations of geothermal development (q. 114), in foreseeing environmental or social problems (q. 116), or in being aware of opposition to geothermal development (q. 118).

Those who feel that citizens of Imperial County have been adequately informed about geothermal development are more likely than those who are unsure or do not believe that they have been adequately informed to know about meetings held to discuss geothermal development (q. 121) and more likely to have visited a geothermal testing site (60% as opposed to about 30%, respectively).

Only four attitudinal statements (qs. 127–149) were statistically

TABLE 6-9. ADEQUATE INFORMATION ABOUT
GEOTHERMAL DEVELOPMENT AND BEING IN FAVOR OR
OPPOSED

	Adequate information (Q. 124)		
In favor or opposed	Yes (%)	No (%)	Unsure (%)
In favor	93.7	90.3	94.1
Don't care	0.0	6.3	5.9
Opposed	6.3	3.4	0.0
Totals:	100.0	100.0	100.0
	(N = 16)	(N = 176)	(N = 68)

significant when residents who felt that citizens of Imperial County were adequately informed were compared to those who did not or were unsure. Those who felt citizens were adequately informed were more likely to agree or be uncertain about the following statement: "Because it will attract new residents, I'm against geothermal development" (q. 130). Similarly, people who felt that citizens were adequately informed were more likely than others to agree that "New developments like geothermal development are not welcome in Imperial County" (q. 137). On the other hand, those who felt that citizens of Imperial County had been adequately informed about geothermal development were more likely than others to agree that "Geothermal development will increase jobs in Imperial County" (q. 143) and to disagree that "Local government officials have primary responsibility to plan geothermal exploration and production (q. 144).

Conclusions

Only a small minority of the Imperial County population feels that they have a very good understanding of geothermal development. Yet, nine out of every ten persons surveyed were in favor of it. There was clear agreement that geothermal development should be strictly regulated. Little local opposition to geothermal development in Imperial County exists and what little opposition there is, apparently is unorganized. Few people had ever attended a meeting about geothermal development. Since such lack of attendance, however, might reflect a paucity of meetings, it is not clear if this response reflects lack of interest among citizens or lack of an informational process. About one of three persons reported visiting a geothermal test site. Very few people feel that citizens of Imperial County had been adequately informed about geothermal development. It should be mentioned, however, that there appears to be a general lack of citizen knowledge in the state and nation about geothermal energy and perhaps about energy in general. Imperial County residents would seem to have greater need for geothermal information since they will be affected by geothermal development to a larger extent. Generally, those in favor of development were also in favor of strict regulation [Freudenberg *et al.* (1977) reported similar findings for a coal boomtown in Colorado].

Regulation of Geothermal Development

Background Characteristics

Almost 90% of the residents in Imperial County believe in strict regulation for geothermal development, which is very similar to residents of Cobb Valley and Lake County, California (Vollintine and Weres, June, 1976). Those who have lived their entire lives in Imperial County are least likely to agree with strict regulation (70%), and the further away a person originally was from Imperial County before moving there, the more likely they were to agree that strict regulation is needed (the range is from 70% to 92% favoring strict regulation). The larger amount of land owned within city limits in Imperial County, the more likely a person was to strongly agree that strict regulation of geothermal development was necessary, although there was no difference by amount of rural land owned.

No differences in the necessity for strict regulation of geothermal development exist by education, occupation, income level, or by the other social demographic data obtained in the questionnaire. Similarly, no difference of opinion about strictness of regulation exists by those who responded early or later to the questionnaire.

Attitudinal Statements

There are no differences in understanding of geothermal development and whether one is in favor of strict or loose regulation. However, porportionately more people who favor little regulation believe that geothermal development would bring cheaper electricity to Imperial County, although regardless of the view toward regulation this was a major expectation of all respondents. Residents who wanted strict regulation also expected geothermal development to bring about beneficial environmental changes. People who were uncertain about strictness of regulation generally expected geothermal development to bring more industry and jobs to Imperial County. Respondents opposed to strict regulation also expected geothermal development to attract more industry and other energy sources to Imperial County. Finally, those who were uncertain in regards to regulation were most likely not to report any positive or negative expectations of geothermal development.

The more in agreement a person was that preserving the environment is an important issue, the more likely he/she was to favor strict regulation. Similarly, they were more likely to foresee social or environmental problems arising out of geothermal development (q. 116). The major problems reported by those who were uncertain about strict regulation were environmental, while those few problems expected by those in favor of no regulation were of a miscellaneous nature, although 90% of those in favor did not respond at all to this question.

There are differences in attitudes about strictness of regulation and awareness in opposition to geothermal development in Imperial County. About equal proportions of people who favor strict regulation and those opposed had attended meetings about geothermal development, whereas *none* of those who were uncertain about regulations had ever attended such a meeting. This might be construed as suggesting that attendance at a meeting concerning geothermal development polarizes attitudes about regulation; however, those in favor of no regulation are more likely than those who are uncertain or want strict regulation to feel that they have an adequate knowledge about geothermal development, although they were no more likely to have visited a geothermal site.

There were no differences in agreement or disagreement about the extent that geothermal development should be strictly regulated and most of the attitudinal statements (qs. 127–149). However, residents who disagreed with the statement that "Economic benefits from geothermal development are more important than environmental costs" (q. 129) were more likely to be for strict regulation than those who were uncertain about strict regulation, who typically were also uncertain in their response to this statement. People who believed in more lax regulation of geothermal development tended to agree with this question. Residents who were both in favor and in opposition to strict regulation were more likely to agree with the statement that "Imperial County can broaden its economic emphasis to more than agriculture through geothermal development" (q. 139) while those who were uncertain about strict regulation were also uncertain in their responses to this statement.

Those who were for lax regulation of geothermal development disagreed with the statement that "Geothermal development may cause unusual odor problems" (q. 141). Those who were uncertain about strictness of regulation or were for strict regulation were uncertain whether or not geothermal development would cause unusual odor problems.

The other statements had statistically significant responses in regard to whether a person was in favor of strict or lax regulation of geothermal development. Those who were in favor of loose regulation were most likely to agree with the statement that "Geothermal companies should have the main responsibility to plan and conduct steam exploration and production" (q. 140), and to disagree with the statement that "Local government officials have primary responsibility to plan geothermal exploration and production" (q. 144), and to disagree with the statement that "Geothermal development will cause border regulations to change making it easier for Mexican National workers to cross into the United States" (q. 149). In contrast to residents who were uncertain or for strict regulation generally responded in an opposite direction to those who were for lax regulation.

Foreseeing Problems Arising out of Geothermal Development

Background Characteristics

Newer residents to Imperial County are more likely to foresee environmental and social problems developing from geothermal development than longer-term residents. Similarly, people planning to spend the rest of their lives in Imperial County are less likely to foresee environmental or social problems arising because of geothermal development. The more rural and city land owned by a respondent, the less likely he/she was to foresee environmental or social problems arising out of geothermal development. Those who own no rural land are most likely to foresee environmental or social problems developing out of it. Nonowners of their dwelling units are more likely than owners to foresee environmental or social problems arising out of geothermal development and proportionately more people who live in duplexes and apartments foresee problems than do residents who live in single-family housing units and mobile homes.

There is a tendency for people with lower *and* higher educational levels, as opposed to those with average levels, to foresee environmental and social problems evolving out of geothermal development; however, occupation, income, and race/ethnicity are unrelated to foreseeing problems.

The earliest respondents to the questionnaire were those least likely to foresee environmental and social problems, and the later one returned the questionnaire, the more likely he/she was to foresee environmental and social problems arising out of geothermal development.

Attitudes

As noted previously, residents who foresee problems arising out of geothermal development are less likely than others to report a good understanding of geothermal development. There was a statistical difference between people who foresee problems and those who do not in regard to being in favor or opposed to geothermal development, yet since most of the population of Imperial County is in favor of geothermal development, the difference is between 88% and 94% who are in favor by foreseeing problems or not, respectively.

Generally, those foreseeing problems from geothermal development were more likely than those who did not to be in favor of strict regulation of geothermal development. While there is a statistically significant difference between those who foresee problems and those who do not, the percentages for strict regulation are 95% and 85.5%, respectively, since the vast majority of the population of Imperial County believes that geothermal development should be strictly regulated.

Those who foresee problems developing out of geothermal development also are more likely than those who do not to believe that property values will increase because of geothermal development. Otherwise, generalized expectations from geothermal development are similar (q. 114). About half of the respondents who did foresee problems reported quite specific ones—for example, one-third mentioned environmental problems, about 10% mentioned a variety of miscellaneous problems, and there were a few people who mentioned population increases as a problem, greed and/or inflation, disaster—e.g., earthquakes—and those who believe that geothermal development will require a raise in taxes because of an increased demand for various kinds of county services. Also, residents who foresee problems are more likely than those who do not to report being aware or unsure about opposition to geothermal development in Imperial County, although there are no consistent patterns reported as to who is in opposition. Apparently, no one knew of organized opposition to geothermal development in Imperial County at the time of the survey.

No statistical differences are apparent between those who foresee problems and those who do not in regard to meetings about geothermal development in Imperial County nor in their feelings about whether or not citizens of Imperial County have been adequately informed about geothermal development (q. 124), nor is there any difference in whether or not they had visited geothermal sites.

The view a respondent had about foreseeing environmental or social problems arising out of geothermal development affected responses to virtually all of the attitudinal statements regarding the effect of geothermal development on Imperial County (qs. 127–149). Response to these statements showed a definite pattern. Persons who felt that there would be environmental or social problems arising out of geothermal development consistently agreed or disagreed with statements that were questioning geothermal development and/or were uncertain about its effects. On the other hand, those people who believed that geothermal development in Imperial County would not result in environmental or social problems consistently responded to these statements in a manner which could be construed as favorable to geothermal development.

In the economic sphere, people who foresee no problems were more likely to agree with the statement that "Geothermal development will bring new tax revenues to Imperial County" (q. 127) while those who saw geothermal development as creating problems were uncertain. Similarly, those who agreed that geothermal development would not pose social or environmental problems were more likely to agree with the statement that "Economic benefits of geothermal development are more important than environmental costs" (q. 129) while those who foresee problems occurring because of geothermal development disagreed with this particular statement. Also, virtually all residents who believe that geothermal development would not create problems in Imperial County agreed that "Because it would attract new businesses and help Imperial Valley grow, I'm in favor of geothermal development" (q. 132). Very few people agreed with the statement that "New developments like geothermal are not welcome in Imperial County" (q. 137), indicating that the vast majority of the population in Imperial Valley and County would welcome new development. Nevertheless, there were differences in the extent of uncertainty in responses to new development when people were cross-classified by whether or not they believed that problems would occur in Imperial County because of geothermal development. Those who felt that there probably would be problems were more likely to respond uncertainly to this statement. Similarly, those who thought that geothermal development would not create problems were somewhat more likely to disagree with the statement. However, the greatest difference was for those who were uncertain about new developments in Imperial County; they were much more likely to report uncertainty about problems arising. Residents who reported that probably there would be problems emerging from geothermal development were more uncertain in regards to the statement "Imperial County can broaden its economic emphasis to more than agriculture through geothermal development" (q. 139), whereas those who did not see any problems emerging from geothermal development were more likely to agree with this statement.

Insofar as potential environmental problems were concerned, residents who felt that geothermal development might create environmental or other problems were more likely to agree with the statement "Noise from geothermal development can be bothersome" (q. 128), and "The construction of geothermal power plants, transmission lines, pipelines and roads which result from geothermal development will create eyesores" (q. 131), and "Geothermal development may cause unusual odor problems" (q. 141). Again, those who foresee problems because of geothermal development were more likely to agree or be uncertain about the impact of geothermal development upon Imperial County in regard to these particular factors. On the other hand, consistently, those persons who did not foresee any problems developing out of geothermal development disagree with these particular statements.

Residents of Imperial County who believe that geothermal development might create environmental or social problems were more likely to be uncertain or disagree with the statement "Because it will attract new residents, I'm against geothermal development" (q. 130). Residents who thought that it would not create a problem were much more likely to disagree with this particular statement. Similarly, both those who thought geothermal development might create problems and those who believed that it would not, substantially agreed with the statement that "Geothermal development will increase jobs in Imperial County" (q. 143). However, those who thought that geothermal development would not create problems were much more likely to agree with the statement that "Geothermal development will cause border regulations to change making it easier for Mexican National workers to cross into the United States" (q. 149). However, of those who did not foresee problems, 77% of them disagreed with this statement.

Residents of Imperial County who foresee environmental or social problems arising out of geothermal development were most likely to disagree with the statement that "Geothermal companies should have the main responsibility to plan and conduct steam exploration and production" (q. 140) and to be about equally distributed on the question of "Local government officials have primary responsibility to plan geothermal exploration and production" (q. 144). On the other hand, those who did not foresee problems arising out of geothermal development were more likely to agree with the statement about geothermal companies having the main responsibility to plan and conduct steam exploration and production and about as equally likely to agree and disagree with the statement about local government officials having the responsibility to plan geothermal exploration and production. Evidently, there is ambivalence about responsibility by those who do not see geothermal development as creating environmental problems.

Residents who foresee problems developing out of geothermal developments were uncertain about whether or not "Geothermal development will increase demands on city and county government and thus increase taxes" (q. 142), whereas those who felt that geothermal development would not create problems substantially disagreed with this particular statement. Residents of Imperial County substantially disagreed with the statement that "Geothermal development will take water away from agriculture" (q. 145). However, of those who foresee problems developing out of geothermal development, 82% reported disagreement, whereas of those who did not foresee problems developing out of geothermal development, only 49% disagreed. Similarly, most Imperial County residents agreed that "The Imperial County policy that new industries, like geothermal, should be able to live with agriculture is a good one" (q. 148). However, of those who foresee problems arising out of geothermal development, 80% agreed with this particular statement, whereas 90% of those who did not foresee any problems agreed with this statement.

Finally, only 8% of Imperial County residents agreed with the statement "I like Imperial Valley the way it is, and don't want it to change" (q. 136). This, of course, is quite likely to be related to the high out-migration from Imperial County that we reported in an earlier chapter. Interestingly, residents who foresee problems developing out of geothermal development were more likely to be uncertain in response to this question, whereas those who do not foresee any problems arising out of geothermal development were more likely to disagree. Two out of every three people foreseeing problems developing out of geothermal development disagreed with this statement whereas 88% of those who did not foresee problems arising out of geothermal development disagreed with it. Thus, while the overall consensus was that Imperial County needed changes, those who thought that geothermal development would not create environmental problems were more likely to be in favor of changes.

Land Ownership and Opinion about Geothermal Development

Land Ownership

Generally it can be assumed that agribusiness dominates the economic, social, and political life of Imperial County. According to an earlier study conducted by Green and Farnan (1975), there is substantial agreement among most leader respondents that the major political and decision-making activity in Imperial County involves large farmers, the Board of Directors of the Imperial Irrigation District, the Board of Supervisors, and certain businessmen. While there are some indications in the Green and Farnan report of a recent broadened decision-making base, very few people of Mexican-American descent actively participate in the major decisions affecting Imperial County. The leadership system described in the Green and Farnan report fits the description of a monolithic leadership network. No one in Imperial County suggested that decisions are not systematically made. It should be noted that this description also fits Imperial County if one uses census bureau data to obtain what is known as a MPO ratio; that is, the ratio of managers, proprietors, and officials to the rest of the labor force. Under these terms, Imperial County also has a strict monolithic structure. The leadership system is explored in more detail in the next chapter.

Among both landowners and nonowners, there are consistent references to the importance that large-scale agriculture has to the county by virtue of its influence on the county's economy. Ultimately, county residents will be faced with another source of impact upon decision making. If geothermal development becomes a large-scale enterprise, consequent large-scale economic inputs into the county's economy by energy companies will become a reality. The potential for conflict will certainly exist, especially over the use of land, water, and services. It cannot be assumed that agribusiness and energy production will always be in concert. If they are not, the potential for conflict over decisions may arise. Thus, if a decision has to be made allocating scarce resources between agriculture or energy, who is to make the decisions and on what basis become extremely important questions to various interest groups.

The potential for Imperial County to evolve into a leadership system different from that which now exists is based on the assumption that decisions are generally made in communities by those persons who represent the functional units that control the flow of sustenance (economic resources) into it. In Imperial County, up until now, this has been primarily agribusiness; however, with the advent of geothermal energy development, energy-producing companies and corporations will be concerned with economic resources at least as great as, if not substantially greater than, those of agribusiness. Thus, the flow of sustenance formerly under the dominating influence of agriculture will then also be under the influence of energy production companies. While there may not be problems while decisions equally affect both interests, the potential for conflict exists when a decision may adversely affect one interest but not the other, or when other scarce resources must be allocated, e.g., water, land, and/or services.

The altered flow of sustenance into Imperial County most probably will result in alternative influence patterns developing within the county as a result of this wealth being able to control other resources, e.g., personnel and community institutions (Butler, 1976 and 1977).

Landowner's Opinion

The landowners of Imperial County generally accept the idea that growth is good for the community. Indeed, this acceptance of the need and desirability for growth permeates all of the responses to the public opinion survey. In this regard, larger landowners and other citizens are in concert. While "environmental protection was found to engender little interest on the part of many community leaders," in Green and Farnan's (1975) study, responses to this public opinion survey by larger landowners clearly show extensive concern with the environment and a desire for strict regulation of geothermal development. This is probably based upon a self-interest of protecting the agricultural environment. On the other hand, there appears to be little interest on the part of larger landowners in developing citizen participation in major decisions, e.g., geothermal development.

This lack of interest in developing citizen input may be reflected by the very few people who responded to the public opinion survey who believed that citizens of Imperial County have been adequately informed about geothermal development. Thus, wider participation by citizens may be viewed as encouraging conflict and controversy that most community leaders, including larger landowners, would like to avoid (Green and Farnan, 1975:33). This is especially important at this time since responses by larger landowners as well as responses by others to the public opinion questionnaire indicate that there are few *overt* social issues currently dividing Imperial County residents. Thus, there are rather consistent attitudinal responses among larger landowners and other citizens in regard to Imperial County affairs and politics.

On the surface, there is consistency in responses of larger landowners and other citizens. This consistency is also substantially so in the political arena as indicated by voting patterns. Yet, there are undertones of potential conflict. Issues have arisen over the past few years that have threatened the general consensus that exists in the county. This potential for conflict could erupt in the future over geothermal development; indeed, while the leadership study reported in the next chapter was being carried out, such an eruption occurred. While citizens are almost universally in favor of geothermal development, they also are almost as extensively in favor of strict regulation. Similarly, only rarely does a person in Imperial County express the opinion that citizens have been adequately informed about geothermal development.

It can be safely assumed that if geothermal development proceeds smoothly, citizens will be satisfied with it. However, if environmental problems or a large-scale accident occurs, probably the latent potential for citizen action against geothermal development and those responsible for it may be utilized by those who are in opposition. Such use of public opinion has been reported in other states by foes of strip-mining and in the development of electric power plants in Montana, Wyoming, the Dakotas, Utah, Colorado, Arizona, and New Mexico (Johnson, 1976).

Rural Land Ownership

Assuming that large landowners are representative of the agribusiness interests in Imperial County, public opinion survey responses are used in this section to compare the attitudes of larger

TABLE 6–10. Rural Land Ownership and Preserving the Environment as Important Issue in Geothermal Development

Preserving the environment as an important issue	Rural land owned (Q. 156)		
	None (%)	1–10 Acres (%)	11+ Acres (%)
Yes	76.6	86.2	60.6
Unsure/no	23.4	13.8	39.4
Totals:	100.0	100.0	100.0
	(N = 197)	(N = 29)	(N = 33)

TABLE 6–11. RURAL LAND OWNERSHIP AND FORESEEING ENVIRONMENTAL OR SOCIAL PROBLEMS ARISING OUT OF GEOTHERMAL DEVELOPMENT

	Rural land owned (Q. 156)		
Foreseeing problems	None (%)	1–10 Acres (%)	11+ Acres (%)
Yes	52.5	48.4	28.1
Unsure/no	47.5	51.6	71.9
Totals:	100.0	100.0	100.0
	(N = 198)	(N = 31)	(N = 32)

landowners to those of other citizens (smaller landowners and non-owners) in regard to potential geothermal development in Imperial County. Our use of the term large for landowners of 11 or more acres does not correspond to the meaning of "large" within Imperial County, where the term often implies landholdings of thousands of acres. However, since the subsample of 11+ acre landholders numbered only 33, adding a larger acreage category to the question would probably have reduced response numbers beyond the limits of significance. The opinions of several "large" landowners are assessed as part of the leadership study in the next chapter.

No differences in reported understanding, being in favor or opposed, or in opinion about strictness of regulation of geothermal and development exist by extent of rural land ownership. That is, whether a person owned no land, had some land, or had a large number of acres is unrelated to opinion about geothermal development.

On the other hand, as shown in Table 6–10, those most likely to believe that preserving the environment is an important issue in geothermal development were the larger landowners (q. 115) while those who owned one to ten acres were least likely to consider the environment as an important issue, albeit a self-interested one.

As shown in Table 6–11, however, large rural landowners were significantly less likely than others to foresee environmental problems arising out of geothermal development (q. 116). Large rural landowners were more likely than others to have attended a meeting about potential geothermal development (q. 122), although they were no more likely to ever have visited a geothermal testing site (q. 126). Also, all rural landowners, as opposed to nonowners, systematically were more likely to report that they had been asked how they felt about geothermal development in Imperial County (q. 125).

Large rural landowners were consistent in being more likely than small landowners or nonowners to believe that geothermal development would bring new tax revenues to Imperial County (q. 127), to believe that noise from geothermal development would *not* be bothersome (q. 128), to agree that the economic benefits of geothermal development are more important than environmental costs (q. 129), and to disagree that it would cause unusual odor problems (q. 141). In contrast, large landowners were less likely than others to agree that geothermal resources in Imperial County should be used for purposes other than electricity, e.g., industry (q. 146). Rural landowners with one to ten acres were more likely than large landowners and nonowners to agree that most geothermal electricity produced in Imperial County would be used there (q. 138).

In summary, if large rural landowners who responsed to this public opinion survey are adequate representatives of agribusiness in Imperial County, large landowners and other citizens are alike in their level of understanding, being in favor or opposed, and in the need for strict requlation of geothermal development. On the other hand, large landowners are more likely than others to view geothermal development in a positive manner. This may be because they are potential lease holders. Similarly, they generally view the economic benefits derived from geothermal development as being more important than environmental costs, which is in contrast to others who are more likely to report being concerned with the environment. Nevertheless, large landowners, who generally are longer-term residents of Imperial County and who plan on spending their entire lives in the county, do view the environment as possibly becoming an important issue in development.

Conclusions. Large landowners and other citizens are in substantial agreement about the beneficial nature but need for strict regulation of geothermal development; however, most landowners and other citizens in Imperial County believe that the public has not been adequately informed about geothermal development. This may be a result of the view of some people that citizen input into the planning process would engender conflict. A counterargument might be that unless citizens are actively involved early in the process, they will begin to question the development and to form and join organizations opposed to it.

Public Opinion in the Heber Area

While our survey had as its goal describing public opinion about geothermal development for a cross-section of all residents of Imperial

Public Opinion about Geothermal Development

County, another survey limited to the Heber area suggests that it may be necessary to be site specific in obtaining such information. For example, Ternes (1978:49) notes that if a careful comparison is made between his Heber site survey and our county-wide one, "a wide discrepancy in the range of data becomes very apparent; leading to this writer's ultimate conclusion that the average resident of Heber has absorbed little or no information and has made little or no impact or provided a representative voice concerning geothermal activity as an issue in this County." Thus, while 19% of the county-wide population felt that they had a good understanding of geothermal development, only 2.6% of the respondents in the Heber survey reported that they had a good understanding of geothermal development; this, clearly, is a major difference.

Similar discrepancies also exist between the county-wide public opinion survey and the localized Heber survey in favoring geothermal development. While we reported that 90% of the county-wide population favors geothermal development, Ternes (1978) argues that this is not so in the Heber area (unfortunately, no data were included in his report on this type of question so we can only rely on his generalized statement). Similarly, more county-wide residents than Heber residents knew about meetings being held to explain or discuss geothermal development, 27.1% to 10.4% respectively (Ternes, 1978:67). In addition, more county-wide residents than Heber area residents had attended meetings about geothermal development (Ternes, 1978:68). Ternes (1978:63) concludes that few Heber residents really care about or have taken part in decisions about geothermal development.

The public opinion survey limited to the Heber area examined the working hypotheses that "few residents of Heber really care about or have taken part (at any level) in the decision to exploit and develop the geothermal resources of the Heber KGRA" (Ternes, 1978:5). The conclusion was "few Heber residents either are aware of or understand fully the implications of various environmental problems that are bound to arise." In addition, the Heber study suggested that the principal promoters and owners of the geothermal facilities to be developed live and work outside the limits of Imperial County. Similarly, there is an argument that after the pilot plant stage, "only a small percentage of the total energy generated geothermally will be available for distribution *within* the Imperial Valley or *to Heber residents.*" Additional arguments are made that the quantity of water required for cooling and injection may have a serious effect on agriculture in the county. Finally, "Heber residents, together with the balance of Imperial Valley's population will receive no reduction in their household electric utility charges either now or at any time after the proposed geothermal plants have come on line and commence generating 'low-cost' power" (Ternes, 1978:6).

The Heber KGRA's population is mainly of Hispanic descent, and they may be out of the mainstream anglo communication channels. In any respect, these highly different findings lead us to be cautious about interpreting findings for larger regions and suggest that future studies must be site specific.

Conclusions and Recommendations

Introduction

This section of the chapter is divided into three major sections: (1) recommendations; (2) a discussion about the transferability of the research methodology; and (3) future sociological research concerned with geothermal development in Imperial County.

Recommendations

The public opinion survey clearly found that most people in Imperial County favored development, although with strict regulation. Nevertheless, we also noted that for site-specific studies there may be some variance, e.g., Heber. However, just as clearly shown was lack of understanding about geothermal development and the extremely few people who felt that the citizens of Imperial County have been adequately informed about it. The following recommendations are made on the assumption that it is desirable to continue having the public in favor of geothermal development:

R1. *Adequate notice of meetings and opportunities for involvement should be provided to all potentially affected citizens.*
R2. *Frequent opportunities for citizen input should be made available at all stages of the planning and development process.*

Also, the lack of understanding by the general public and the felt need for information about geothermal development suggests that more formalized informational and educational programs need to be developed and made available throughout the county. This recommendation assumes that an informed public and citizenry is less likely than an uninformed one to react negatively to geothermal production when large-scale development takes place in the future.

The process of actively involving the public assumes the following (Armstrong and Butler, 1974:24–26):

1. People have the right to participate in decisions that affect their well being.
2. Participatory democracy is a useful method for conducting community affairs.
3. People have the right to strive to create the environment they desire.
4. People have the right to reject an externally imposed environment.
5. Maximizing human interaction in a community will increase the potential for humane community development.
6. Effective community planning assists human beings in meeting and dealing with their environment.

Fostering citizen participation also probably would make the planning process one more actively (1) controlling events and delineating viable options and maximizing future options and alternatives and (2) minimizing future uncertainty for the public, both of which should lead to a continued positive stance toward geothermal development.

Transferability of Research

The research methodology used in the public opinion survey can be utilized in any community in the U.S., for any kind of energy development. The extent to which the substantive findings actually apply to other areas is problematic, but the standard research methodology used in this research can be used elsewhere.

Future Research

The population of Imperial County generally, at this time, believes that geothermal development in Imperial County will be beneficial without doing serious damage to the environment and with virtually no social costs. These views, of course, must be tempered with the knowledge that only one out of five persons in Imperial County feels that he or she adequately understands geothermal development. Further, the questionnaire focused on public perception of possible environmental degradation rather than on public commitment to achieving environmental quality.

One next research step would be to measure public preferences for allocating generated tax funds from geothermal development among various expenditure areas—including environmental quality (Dunlap and Dillman, 1976). This is a shift from pregeothermal development as a source engendering possible social and environmental problems to postdevelopment and what should be done with tax revenues generated by geothermal development.

While several of our associated research components have stressed the probability that geothermal development will have relatively little population, employment, and economic impact upon Imperial County (see earlier chapters), it should be recognized that in other areas that have undergone various kinds of development the input of new economic resources has drastically influenced the local area in a variety of ways. The potential exists for large-scale impacts in the development of industry and in-migration of a large population. If industry moved into Imperial County on a large-scale basis, it would bring a population with life-styles essentially different from those now existing in the county. Subsequent population growth would trigger physical growth in the towns and former rural areas requiring land use changes, additional services, and expanded community institutions such as in the political arena, education, religion, recreation, and so on. Thus, there needs to be at least some considered development and evaluation of possible alternative growth patterns.

Finally, the large-scale development of geothermal resources in Imperial County will result in at least some changed land utilization patterns if past research can be used as a reliable guide. The flow of large-scale economic resources generated by geothermal development and its aftermath has to involve major decisions. To what extent conflict will be engendered between the "old" and "new" economic interests will be of utmost importance to the current landowners and citizens of Imperial County. How conflicts are ameliorated will influence the county for coming generations. Research along these lines is essential if current citizens are to continue having a major impact in the county.

Acknowledgment

This chapter was derived from the report of the "sociological" component of the NSF/ERDA funded research titled *Planning For Resource Development: Geothermal Energy in Imperial County, California.*

7

Leadership, Community Decisions, and Geothermal Energy Development: Imperial County, California

INTRODUCTION

Power structures are made up of decision makers who are largely responsible for the actions and nonactions in organizations at all levels in the United States. At the individual level, the ability to make decisions enables one to influence the behavior of another. At the communty-system level, an organization or power group may be able to command the behavior of other individuals or organizations. From a social system point of view, decisions involve every unit of human organization: the individual, the family, voluntary as opposed to involuntary organizations, the government, corporations, and the community (Hawley, 1971). Power is obtained by controlling that which is valued by people in a society (Lasswell and Kaplan, 1950). In the U.S., those who control economic institutions have power, influence decisions, and can implement decisions (Goldberg and Lindstromberg, 1966). Power structures are defined as the characteristic pattern within a community whereby resources are mobilized and sanctions employed in making decisions (Walton, 1967). Thus, a community is considered an organization of units held together through the use of power.

Social science research indicates that influential community leaders usually control important economic and governmental positions, resources, and decisions. The Imperial County economy is currently dominated by agriculture, which comprises about 70% of the total output of the country and 37% of the employment (Lofting, 1977). Agricultural development in the county is characteristically very capital intensive and productive compared with other areas of the county. Economic forecasts, though, indicate that the geothermal industry sector may grow to equal the agricultural sector of the Imperial County economy (Lofting, 1977).

If influence is closely related to the important economic sectors and if geothermal and related industry becomes as large as agriculture, the leadership structure may change substantially. Decisions, once made with relatively little conflict by a small group of leaders having a unified power base and economic interests, may in the future be made by a more diverse power group having conflicting interests competing for scarce resources such as water, land, or labor.

The purpose of this part of the Imperial County research was to delineate the power structure operating in Imperial County, the influential leaders, the source of their power, their likely reactions to geothermal development, and the possible effects geothermal development will have on the power structure itself.

We performed a survey research analysis to elicit opinions of the leadership and the public concerning geothermal development in Imperial County. This analysis indicates the reaction of the power structure to emerging geothermal resource development. The power-structure analysis, combined with forecasts of the economic effects of geothermal development, indicates the ways in which the power structure itself may change. In the remainder of this chapter we examine power and its utilization, the power structure, influential people, decision making, and the impact of these on the people in Imperial County, California.

THE POWER STRUCTURE IN IMPERIAL COUNTY

Methodology

The power structure in Imperial County was identified by using a combination of research methodologies. First, we compiled a list of people who held important positions in Imperial County. This list included people holding positions in government, quasigovernment, business, agriculture, and various associations. Each of these persons, by virtue of their positions, was assumed knowledgeable about at least some issues requiring decision making in Imperial County. The list included a random selection of business enterprises in the county. A representative of all business and agricultural enterprises with 50 or more employees was also included.

A selection of people from this list were interviewed. Part of the interview schedule included questions about people whom the respondent considered the most influential in Imperial County, their occupational and other important positions in the community (e.g., lawyer, charity official, mayor, department store owner, etc.), the extent of their influence, and the basis or source of the person's influence. In addition, other questions were asked about various issues, including geothermal resource development.

As interviews were completed, a card-filing system was used to determine those having a reputation for leadership and influence in Imperial County. A few names emerged that were not on the original position list. These names, obtained from the interviews, were added to those to be interviewed. Because only a few names were added in this manner, we assumed that all the important leaders of Imperial County were known to us. This assumption was validated because no additional names of influential people were added in all of the subsequent interviews. A total of 105 interviews were conducted in 1977 and 1978 from the final list that was compiled by using positional analysis and names added by subsequent interviews.

In summary, the power structure of Imperial County was determined in a systematic way, combining methodologies used in previous research. People who we assumed were influential were identified for an interview on the basis of positions they held in the community.

Influential People and the Power Structure in Imperial Valley

Who is influential in Imperial County? Because agriculture is the dominant economic activity in the county, it is not surprising that many influential people are involved in agricultural pursuits. However, the influence among leaders in the county differs substantially, and not all individuals who have power are directly linked to agriculture. Nevertheless, our evaluation of the power structure suggests that it is, in fact, monolithic, i.e., established, repetitive, and predictable patterns of decisions are made by a rather small group of people in Imperial County.

Clearly, most of the influential leaders in Imperial County combine the two major aspects of power as set forth by Weber (1957): (1) personal attributes and (2) power as part of established authority, in this case resulting from agricultural, governmental or quasigovernmental positions, and, at times, jointly held positions in both spheres.

From our analysis of Imperial County, two people are far above all others in influence and can be considered the dominant, influential people in Imperial County. AG-11,† the most dominant and influential, has large-scale agricultural interests and also has an important government position. The second most influential person, M-11, does not hold a government position and is one of the few important, influential people in Imperial County having no known direct link to agriculture.

Four other people make up the first echelon of leadership in the county. Three of these influential people are directly linked to various kinds of substantial agricultural enterprises (AF-11, AF-12, AF-13). The fourth person is a local businessman who also holds an important government position; he is said to represent the Mexican-American community (BG-11).

Thus, four out of six of the major leaders in Imperial County are directly linked to significant agricultural enterprises. In addition, two of these four hold important governmental positions. Two of the six have no direct relationship to agriculture but by virtue of their positions can greatly influence decisions related to agriculture as well as virtually all other decisions in the county.

A second level of leadership, consisting of nine people, has less influence. Only four of these leaders are directly linked to agriculture (AG-22, AG-24, AG-28, A-29). Again, three of these four, in addition, hold important governmental or quasigovernmental positions. The remaining five second-level influential people all have important governmental positions, and several of them also own large businesses.

A third level, about as influential as the second, contains four people. Three of these four are heavily involved in agriculture. In addition, two of them have or have had important government posts. The remaining person in this group is a local businessman with no apparent ties to agriculture or to any governmental or quasigovernmental position. He is one of the few leaders who is considered a political activist.

Another level of leadership in the county consists of people who hold a variety of positions, some of which, on the surface, seem to be very important and some not so important. These 11 influential people are probably effectors. Effectors are those who put policies decided by others into action. Eight of these 11 influential people hold or have held important governmental positions. The positions of several of these people overlap substantially, however, in both agriculture and

†This is a code number for use in further discussion.

Leadership, Community Decisions, and Geothermal Energy Development 239

business (BG-41, AG-43, AG-44, MG-45, AG-46, and BG-491). Two of these effectors hold only governmental positions (G-42 and G-47).

The interviews of those people who were reported to be the most important leaders by others in Imperial County were analyzed further. If only the key influential people had been interviewed, the results would have been as indicated but with a somewhat stronger demarcation between the first six leaders and all of the others. Furthermore, not one lower-level effector was mentioned by the upper-level leaders as having influence.

Some respondents refused to name specific influential individuals, but many of these same people reported influential groups. In descending order of perceived importance, these groups were farmers with large land holdings, the county supervisors, the councils of each city, minority coalitions, the news media, and the Farm Bureau head. In addition, a large number of other people and positions were reported by key influential people as having power.

Among all of those mentioned by name, only one of the upper-level leaders and one of the effectors had a reported meaningful link to the minority community. Similarly, no one who could be considered an activist (a person who lacks an institutional power base) was reported to have influence or to be an effector in Imperial County. Only one person was reported by several people to be an activist in the county but was considered generally to be ineffective, except as an agitator. One third-level leader was reported to be influential because he was active in political affairs, although he did not hold a political office himself.

Summary

Among the top six leaders who are perceived to have the most influence, two have far more influence than the others. Most of these key leaders are involved in agriculture and also hold important governmental or quasigovernmental positions. Other influential leaders and effectors also seem to be dominated by those with a link either with government or with agriculture. One of the two most influential leaders in the county is an exception in that he is involved neither in agriculture nor in government. However, this leader's position is such that he can influence agricultural, business, and virtually all other decisions that affect the county.

Our analysis, we should note, agrees partially with that of Green and Farnan (1975). They noted that "There is agreement that the most significant political activity revolved around an elite group which consisted of farmers with large land holdings, the Board of Directors of The Imperial Irrigation District (IID), the Board of Supervisors, and certain businessmen in the area." They further suggested that areawide decision making is becoming more representative of the broader community. Our analysis, performed in 1978, clearly indicates that the latter was not true then, and the former is only partially correct in that certain owners of large farms and/or their representatives dominate the decision making. However, not all board members of the IID and the Board of Supervisors have equal power, and some strict qualifications must be made about the influence of local businessmen.†

Leadership Opinion and Reaction to Geothermal Development

In this section of the chapter we evaluate opinions of leaders in Imperial County about geothermal resource development. About 90% of the leaders in Imperial County believe that geothermal development is very important and of immediate concern for the county. Only 1% believe that current energy is adequate; 9% believe that geothermal development is important but not of immediate concern. Around 80% of the leaders strongly favor geothermal development in the county, the remaining 20% are in favor, but voice several qualifications, such as "as long as it doesn't harm agriculture" or "if oil companies are closely regulated." Not one leader interviewed was opposed to geothermal resource development (see Green and Farnan, 1975, for similar results).

When questioned about regulation, well over half of the leaders expressed strong opinions that geothermal development should be strictly regulated, another 30% believed less strongly that regulations should be imposed, 11% were uncertain, and 3% believed that no regulation should be imposed. Thus, more than 80% of the leaders felt that geothermal development should be strictly regulated. This question elicited, in addition, a variety of comments. The most prevalent comment was that strict regulation was the only way to avoid problems such as adverse effects on agriculture, subsidence, and monopoly of the resources by oil companies. A substantial belief also exists that it should be strictly regulated because geothermal resources should be

†Events in 1979, since the survey work was completed, may have altered the leadership structure somewhat. The newly elected Board of Supervisors and IID Board have fewer members with ties to agricultural interests than previously. The impact of these changes is not yet clear.

viewed as a public utility or a resource belonging to everyone. Generally, the volunteered comments reflected a great deal of knowledge about geothermal development by some of these leaders.

According to almost half of the leaders, the oil companies are primarily responsible for initiating geothermal development in Imperial County; another 25%, private enterprise; others, Dr. Rex and/or the University of California, Riverside; and a small number, the IID or the Magma Power Company. One or two others listed a variety of extra-local (federal government, San Diego Gas and Electric Company, Department of Energy, etc.) and local (Board of Supervisors, Public Works Director, local government, etc.) groups as being most responsible.

Almost a third of the leaders believe that opposition to geothermal resource development exists in Imperial County. No one was able or willing, however, to pinpoint a specific individual or group who was opposed. A few believe that San Diego Gas and Electric Company and nuclear power interests are opposed, and some say that the Farm Bureau and unspecified agricultural interests are opposed.

In responding to a question asking for "comments about geothermal development in Imperial County that we didn't discuss and you feel we should have," the major responses were: (1) too many government regulations exist, (2) development has been too slow, (3) all levels of government should be involved in geothermal development, and (4) more education and/or information should be made available to the general public. The only other numerically meaningful responses were that the federal government, *a la* the Tennessee Valley Authority, should control geothermal development. In contrast, many leaders believed that the local county government should control it.

A Comparison of Leadership Opinion and Public Opinion of Geothermal Resource Development

Table 7-1 shows the opinion of the public, leaders in general, and the top fifteen leaders on various aspects of geothermal development in Imperial County. In comparing the key questions in Table 7-1, it is apparent that leaders are generally more in favor of geothermal development than the general public. Leaders are also slightly more likely than the general public to believe that geothermal development may create some problems in Imperial County.

Leaders are less likely than the general public to believe that economic benefits are more important than environmental costs (q. 129).

TABLE 7-1. OPINIONS ON GEOTHERMAL RESOURCE DEVELOPMENT IN IMPERIAL COUNTY, CALIFORNIA, 1977-1978.[a]

	General public opinion[a] (%)	Strongly agree/agree		
Statements[b]		All leaders (%)	Top 15 leaders (%)	Other leaders (%)
127. Geothermal development will bring new tax revenues to Imperial County.(+)	66.2	94.2	93.3	94.4
128. Noise from geothermal development can be bothersome.(−)	13.0	13.5	33.3[c]	10.1
129. Economic benefits from geothermal development are more important than environmental costs.	37.2	21.2	20.0	20.3
130. Because it will attract new residents, I'm against geothermal development.(−)	3.7	1.9	6.7	1.1
131. The construction of geothermal power plants, transmission lines, pipelines and roads that result will create eyesores.(−)	13.8	23.1	20.0	23.6
132. Because it will attract new businesses and help Imperial Valley grow, I'm in favor of geothermal development.	75.1	90.4	86.6	91.0
133. Most geothermal electricity produced in Imperial County should be used in Imperial County.	52.8	47.1	53.3	46.1
134. A fuel shortage will develop in the United States unless geothermal and other sources of energy are developed.(+)	73.6	82.7	86.7	82.0
135. Geothermal energy will provide cheap electricity for Imperial Valley.	42.4	33.7	26.0	34.8
136. I like Imperial Valley the way it is, and don't want it to change.	7.4	15.4	0.0	7.8
137. New developments like geothermal are not welcome in Imperial County.(−)	5.9	6.7	0.0	7.8
138. Most geothermal electricity produced in Imperial County will be used in Imperial County.	22.7	12.5	20.0	11.2
139. Imperial County can broaden its economic emphasis to more agri-				

(cont.)

	General public opinion[a] (%)	Strongly agree/agree		
Statements[b]		All leaders (%)	Top 15 leaders (%)	Other leaders (%)
culture through geothermal development.(+)	73.6	87.5	80.0	88.7
140. Geothermal companies should have the main responsibility to plan and conduct steam exploration and production.	53.9	69.2	60.0	70.8
141. Geothermal development may cause unusual odor problems.(−)	8.9	18.3	6.7[c]	3.3
142. Geothermal development will increase demands on city and county government and thus increase taxes.(−)	16.0	11.5	6.7[c]	12.3
143. Geothermal development will increase jobs in Imperial County.(+)	81.0	91.3	86.7	91.1
144. Local government officials have primary responsibility to plan geothermal exploration and production.	33.5	50.0	73.4[c]	46.0
145. Geothermal development will take water away from agriculture.(−)	4.8	9.6	33.3[c]	5.6
146. Geothermal resources in Imperial Valley should be used for purposes other than electricity, such as by industry or for chemicals.	39.4	55.8	73.3	52.8
147. Geothermal development will result in fewer Mexican National agricultural workers crossing daily into Imperial Valley.	7.1	24.0	20.0	24.7
148. The Imperial Valley policy that new industries, like geothermal, should be able to live with agriculture is a good one.(+)	82.2	93.3	93.3	93.3
149. Geothermal development will cause border regulations to change, making it easier for Mexican National workers to cross into the United States.	5.9	1.9	0.0	2.2

[a] Questions 1–126 refer to an earlier survey of the general public in Imperial County taken in 1976 (Butler and Pick, 1977).
[b] (+) Clearly in favor of geothermal resource development; (−) clearly against geothermal resource development.
[c] Statistically significant difference between top leaders and other leaders.

Also, a greater percentage of the general public than of the leaders believes that geothermal energy development will provide cheap electricity for local residents and that most of the locally produced electricity will be used locally.

Leaders, more than the general public, would give geothermal energy companies the main responsibility to plan and conduct exploration and production. On the other hand, more leaders than the general public would give local government officials the primary responsibility.

Leaders, more than the general public, believe that geothermal resources should be used for nonelectrical purposes. More of the leaders also believe that such development will reduce the number of Mexican workers in the county.

In comparing responses of the top leaders and other leaders, few major differences in opinion were noted. More of the top leaders, however, believe that noise from geothermal development might be bothersome. In fact, a similar percentage of other leaders and the general public believe that geothermal development will increase demands on city and county government and thus increase taxes; fewer top leaders believe geothermal development will have these effects.

More of the top leaders, again in contrast to other leaders and the general public, believe that geothermal development will reduce the availability of water for agriculture.

Although these differences exist, top leaders and other leaders hold substantially the same opinions on all the other statements on various facets of geothermal development.

The Effect of Geothermal Resource Development on the Power Structure in Imperial County

What is the likely impact of geothermal resource development on the leadership structure in Imperial County? Several research studies have concluded that geothermal energy development will probably create substantial impact on the population (Pick, 1977), employment and the economy (Lofting, 1977), and the fiscal system (Goldman and Strong, 1977) of Imperial County. However, these anticipated social and economic effects are, by and large, beneficial. The few negative ones are expected to be relatively small and manageable. Other impacts, including those on water quality, air quality, aquatic and terrestrial biology, health, and seismicity are generally negligible (Layton, 1979). Exceptions include cooling tower drift, accidental release of brine, and subsidence. Even these potential problems, though, can probably be managed. Some of the mitigation efforts will require substantial capital investment and application of technology and skilled labor. However, the point is that geothermal resources can generally be developed compatibly with existing agricultural activities. The opinion research reported here demonstrates that the perception of Imperial County leaders of the effects of geothermal development on agriculture agrees with the technical research.

However, leadership patterns have been drastically changed in other regions affected by energy development. If industry moves into Imperial County on a large scale, a population with essentially different life-styles from the people now living in the county will immigrate there. Subsequent population growth will trigger physical growth in the towns and former rural areas, requiring land-use changes, additional services, and expanded community institutions, such as administration, education, religion, recreation, and others. The flow of such large-scale economic resources involves major decisions. The development of geothermal resources in Imperial County will result in new leadership and influence patterns, if past research is a reliable guide. To what extent conflict will be engendered between the old and new economic interests and how conflicts are resolved will be vitally important to the leaders and citizens and will influence the county for coming generations.

The current leadership in Imperial County—unanimously supportive of geothermal resource development, though advocating strong controls—apparently are confident that geothermal resources can be developed without threatening agriculture, the current dominant economic base of the power structure. It can be assumed that the leadership has already responded positively to geothermal development by supporting it privately through lease agreements (a necessary condition for the current level of exploration and experimentation) and publicly by influencing county policy. Strong controls have been built into this support through the Geothermal Element of the General Plan, environmental review, and use conditions. The controls are primarily directed toward minimizing damaging conflicts between geothermal resource development and the existing agricultural development, with secondary concern for other environmental issues.

Influential leaders in Imperial County support controlled development of geothermal resources because they believe that new industry will not threaten continued agricultural activities and revenues; they believe that geothermal development will yield additional revenues in the form of leasehold and royalty payments. In other words, the lands controlled by current influential leaders have two sources of revenue, one from surface agricultural use and the other from subsurface geothermal resource development, without serious interference between them. The leaders of Imperial County believe that their economic position will not be threatened, but enhanced.

The power structure may be affected by the needs for outside capital, for managerial and technical expertise, and for a strong, constant, secure market in the form of an electric utility. The needs of these outside interests will have to be met so that the indigenous power structure can realize revenues from geothermal resource development. This change will probably take the form of cooperative accommodation, rather than outright sharing of or a change in the locally based power structure, for two reasons: First, it will be in the interests of the geothermal developers and the utility to disturb the current power structure as little as possible to facilitate development. Second, the vast majority of the land from which the geothermal resource will be extracted will remain in the hands of owners who will continue current surface uses.

One major potential conflict that may develop between the surface agricultural interests and the new subsurface resource developers is over scarce water supplies. Agriculture in the valley uses significant quantities of water (3 million acre ft/yr) to irrigate crops (up to five plantings a year) and to leach salts from the soil. Heat exchangers used in geothermal power plants will also require substantial quantities of cooling water. Current analyses of the lower Colorado River basin hydrology, legal constraints, geothermal technology, and Imperial County policy indicate that county-wide geothermal development will have few constraints up to 7000 MW_e of generating capacity (Layton, 1979). However, specific subareas of the county may have water shortages before the 7000 MW_e county-wide capacity is developed.

Whether or not there is a conflict, and to what degree, depends on a number of factors difficult to determine at this time. These variables include: the heat-exchange technology used; the success of reinjection; the availability, at an economically feasible cost, of treated agricultural drainwater; the rate of and total extent of geothermal resource use; the types of crops planted; the extent and success of water conservation efforts; the status of upstream claimants to Colorado River water; basin hydrologic performances; IID policy with regard to irrigation water; and county policy.

Conclusions

Power is the ability to command the performance of individuals, groups, and organizations. A systematic, patterned use of power exists in Imperial County, structured as a monolithic leadership system. This monolithic structure, not too surprisingly, is dominated by agricultural interests, although one of the two most influential leaders in Imperial County is not directly linked to agriculture His position, however, allows him to influence agriculturally related decisions. Agricultural interests in the county are systematically interlocked with local government, i.e., many of the influential leaders have large-scale agricultural enterprises and also hold important local governmental or quasigovernmental positions, some elected and some appointed.

The leadership in Imperial County is a visible one. However, the power and influence of individuals in Imperial County varies substantially, even though they ostensibly occupy the same or similar positions. Similarly, some individuals who hold positions that, on the surface would seem to give them power, do not actually have extensive power. Some of these individuals are not even considered to be effectors or lower-level influential leaders; these people could be considered symbolic leaders because some citizens and outsiders assume they are influential, but other leaders, especially the top ones, do not consider them influential in important decisions.

This research and most previous studies illustrate the importance of personal attributes, in addition to positional authority, in power and influence in the community. Wealth alone is a poor indicator of power in Imperial County. Yet, almost all of the key leaders control substantial economic resources, especially in agriculture.

Key leaders know who the others are, systematically list them, and do so much more often than do less influential citizens. A substantial consensus seems to exist among key leaders on most issues, although they may differ slightly on the implementation of decisions or on minor issues. As far as geothermal resource development is concerned, they all are in favor of it, but most of them want strict regulation.

A strong consensus exists among other leaders' appraisal of key leaders, key leaders' self-evaluation, and their actual influence in Imperial County: a small group of individuals influence all of the major decisions in the county. Their influence cuts across all issues, including geothermal development. Clear, structured, purposeful decision making occurs, and the decisions affect the lives of all the citizens in the county.

New population, attracted by geothermal energy and other commercial and industrial development, will probably have characteristics, life-styles, and demands for community services different from those of existing residents. However, the leadership structure will probably not be significantly affected. Surface agricultural use, currently the dominant economic sector, is generally compatible with the extraction and conversion of subsurface geothermal resources to electric or direct heat energy. The influential leaders, already in control of substantial agribusiness revenues, will derive additional revenues in the form of geothermal lease and royalty payments from the land resources they control. Competition between agricultural needs for irrigation water and electrical production needs for cooling water will become a problem county-wide only if electrical production reaches high levels. Certain subareas of the county could have water shortages at lower electrical production levels as a result of the distribution of water, irrigation systems, and power plants. The development and extent of competition for water depends on a number of factors that will not be resolved for some time.

Acknowledgment

Parts of this chapter were published by Lawrence Livermore Laboratory as the Final Report for contract No. 4679303 (W-7405-ENG-48). Mr. Charles Hall of LLL was instrumental in carrying out this research and in writing this chapter.

8

Geothermal Development Update

INTRODUCTION

The socioeconomic characteristics and geophysical parameters for Imperial County have undergone detailed exploration in earlier chapters. The unfolding of geothermal development in a real community, however, is not an orderly process. The manner in which the county actually will be impacted by geothermal development depends partly on chance combinations of local and extralocal decision making and events. The goal of this chapter is to present the chain of recent events in Imperial County's geothermal development process. With knowledge of the latest geothermal events, the policy and research recommendations presented in Chapters 1 and 9 can be viewed in the context of what has happened so far in the development process.

The county's pace of change in geothermal development has been rapid. For instance, federal and state government decisions have caused rapid and often unanticipated chains of events. One such decision was the alteration in geothermal taxation laws in the National Energy Bill in the Fall of 1978. The passing into law of the investment tax credit and depletion allowance for geothermal resources has led to a rapid chain of events in the financing area. The pace of change will continue to be rapid. As development unfolds, policy makers, companies, and investors concerned with Imperial County will need continually to seek the fullest and most recent updated information.

Chapters 3 through 7 focused on the county population, public opinion, and leadership features relevant to prospective geothermal development. The primary concern of this chapter is to identify current events marking the first events of the actual geothermal development process in the population, public opinion, and leadership areas. The chapter then examines current geothermal projects in the county. These are grouped into electrical and direct use categories. Attention is next focused on current geothermal regulatory hurdles in the county. Such regulations have a major influence on the length of time for implementation of geothermal projects. The last part of the chapter examines limiting factors on geothermal development in the county, including transmission corridors, financing of projects, transmission pipes, and cooling water.

Beginnings of Geothermal Impact on County Population and Leadership

Earlier chapters outlined the potential effects of geothermal development on population, citizen opinion, and leadership. Early signs of the development process are now detectable in the population and leadership areas. On the other hand, in the public opinion area, there appear to be few signs of change because of geothermal development.

Why has public opinion not reflected geothermal issues? There appear to be two reasons? First, there is a reluctance by county leaders to make initial geothermal issues public, although, as discussed below, several geothermal issues have been debated by leaders. Second, even if issues were under active debate by the citizenry, it is difficult in such a small rural county to detect and measure public issues without a major political election and/or a formal public opinion survey, like the one reported in Chapter 6.

In the population area, initial impacts appear limited to migration into the county of small numbers of energy company workers and their dependents. For example, at the start of 1979, Westec Services,Inc., a geothermal contractor and consulting company, had about 60 workers located in the county. About half were county residents trained locally. The remainder were Westec employees, largely managers and professionals relocated from the San Diego area. The impact of 30 in-migrating workers to a county labor force of 38,525 is minimal. However, there is greater impact if only the skilled employment sector, professional/managerial, is considered. In 1970 this sector numbered 4,889 so that the addition of 30 Westec workers represents about a 0.6% sector increase. The county labor force is low-skilled by statewide trends. Thus, the training in geothermal job skills of the 30 locally recruited Westec workers is important relative to a small skilled sector. As the advent of production in the North Brawley geothermal field has drawn closer, Union Oil Company has increased the number of employees located in Brawley from 5 to about 50. In Union's case, nearly all workers are in-migrants who are skilled workers able to uti-

Geothermal Development Update

FIGURE 8-1. Geothermal pipe, Union Oil drillsite, North Brawley.

lize the equipment necessary to drill the wells, Figure 8-1. Most are either professional/managerial or highly skilled. When the employees of other geothermal companies, such as Magma Power, Republic Geothermal, and Chevron Resources are considered, a total of several hundred skilled in-migrant workers and dependents have so far migrated into the county for geothermal development reasons.

Will most of these workers leave the county after geothermal production is achieved? The answer relates partly to alternative power plant capacity sequences, one of which was shown in Figure 5-3. If plant capacity is rapidly increased, such workers may be maintained by their employers as longer-term county residents.

Several incipient geothermal developments have involved county leadership, including a dispute over the location of the Brawley sewage plant, conflict over the union affiliation of geothermal labor, and conflict over the routing of a high-voltage transmission corridor. The origin of the sewage plant dispute antecedes the decision by Union Oil to develop the North Brawley geothermal field. The City of Brawley built a crude primary sewage plant in the 1950s. Because of lack of regulation, wastes could be dumped directly into the New River. By chance, the sewage plant location happened to coincide exactly with what subsequently has become a prime area for geothermal development. Several years ago, a plan was established by the City of Brawley to put in a $3 million secondary sewage plant with ponding and aeration in place of the previous primary treatment. It was also the intent of the City Plan to use treated sewage effluent as fertilizer in the farming of nonedible crops. For this purpose, an action was started by the City of Brawley with respect to 400 acres of land near the proposed sewage plant. This was based on an economic analysis, which included the alternative of secondary treatment. The city favored doing the farming itself. Another city alternative of installing more advanced secondary sewage treatment was ruled out because of equipment depreciation and maintainance costs. The parties of the dispute are the landowners of the land proposed for condemnation and the city. The dispute almost reached the stage of legal steps in a San Diego courtroom. The landowners sought to regain ownership of this prime land, while the city hoped to realize its original plan, regardless of geothermal potential. One pending solution involves Southern California Edison Company's using the effluent as make-up water for geothermal cooling purposes. The lawsuit is not being pursued, but if SCE's use of the make-up water falls through, the legal issue will be reviewed again. Meanwhile, the Regional Water Quality Board has laid out a timetable for a firm contractual agreement to use the effluent for geothermal purposes. For the present, the city plans to put the effluent in holding ponds, with a 30-day water supply, which are necessary no matter which way the dispute is resolved.

Another leadership dispute surfacing in 1978 involved union affiliation of geothermal workers. Although geothermal workers have special training requirements, geothermal power contracts involving at least one utility, San Diego Gas and Electric Company, require geothermal workers to belong to the International Brotherhood of Electrical Workers. The dispute involved a vocational training program for geothermal workers which was initiated by CETA, under DOE sponsorship. This program was conducted by Westec Services, Inc., in conjunction with Imperial Valley College. Its goal was to train ethnic minorities and other economically disadvantaged workers in geothermal job skills. Such a goal coincides with one stated in the Geothermal Element of the Imperial County General Plan. This dispute has involved the question of which group—union, or disadvantaged minorities—should be favored in filling geothermal jobs. As of January, 1979, Westec Services had withdrawn as job trainer, leaving a stalled training program. Nearly all local geothermal jobs are being filled by union workers.

A third geothermal-related leadership dispute has developed over the location of transmission line corridors. This conflict is of greater significance to the county's geothermal development process than the two above disputes. In the latter part of this chapter, the transmission corridor factor is considered to be the most important limiting factor on geothermal development.

Why are the corridors so important? An existing 640-mile network of IID transmission corridors already is in place to transmit the fossil and hydroelectric energy produced from the existing 350 MW_e of IID energy capacity. The existing transmission lines in 1976 are shown in Figure 8-2. These consist of 298 miles of 161-kV (kilovolt) lines, located mostly in the central and eastern parts of the county, and 340 miles of 91-kV lines, located primarily in the west. Converting voltage units into capacity units, the 161-kV lines can carry electricity for about 170 MW_e. In the north, the existing network can tap into the existing Southern California Edison (SCE) lines to Los Angeles and, in the east, into Arizona's power grid.

An easy solution appears to be merely widening existing transmission corridors. Seventy-seven percent of the present corridor network has a fixed width of 50 feet, with the remainder of the network occupying corridors of 90–100 foot fixed width, or county roads, irrigation canals, railroad easements, IID property, or Bureau of Reclamation property. However, a 500-kV corridor requires a width of 800 feet at the base of a transmission tower. The requirement of an over ten-fold width increase means that existing rights-of-way are of little use for high-voltage lines. These latter lines must involve new rights-of-way with the attendant problems of environmental impact and land condemnation. A planning corridor is defined as an area wider than the tower bases, which is wide enough to ensure adequate space for corridors given irregular land use and environmental and agricultural restrictions. A recent study (Imperial Valley Action Plan, 1978) recommends a planning corridor of 5 miles width for twinned 500-kV lines and 1 mile width for twinned 230-kV lines.

A stark fact of the existing network is that even under the best of circumstances, a maximum of about 300 MW_e of geothermal electricity can be transported out of the county into either Los Angeles or Arizona. Under realistic conditions, such a maximum would likely be lower, because of the transmission requirements of present fossil- and hydro-generated electricity over the network.

In terms of power plants, a maximum of only six standard 50-MW_e geothermal plants can be accommodated in the Imperial Valley without the construction of new power corridors and transmission lines.

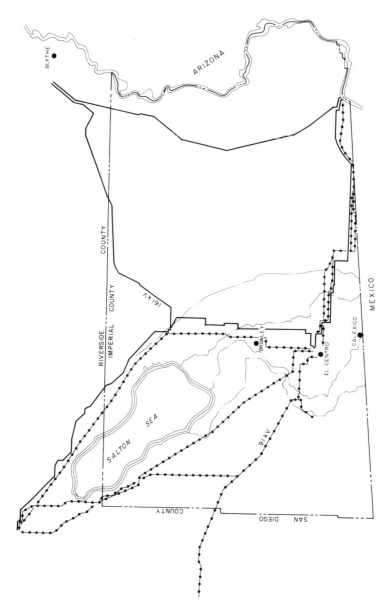

FIGURE 8-2. Existing IID network of transmission lines. These lines, of either 91 or 161 kV capacity, presently provide Imperial Valley consumers with fossil- and hydro-produced electricity. (Source: Imperial Valley Action Plan, 1978).

Geothermal Development Update

The Jet Propulsion Laboratory medium growth scenario of electrical generating capacity, presented in Figure 5-3, exceeds 300 MW$_e$ in its first five years, and rises to a 3500-MW$_e$ level over a 30-year period. Thus, for this average scenario, a limiting factor occurring almost immediately is construction of transmission lines and widened corridors.

From its beginning, the geothermal transmission corridor dispute has been linked with related disputes over nuclear energy. The fundamental cause of this linkage is the small amount of demonstrated geothermal capacity and geothermal energy's small unit size compared to nuclear facilities. It is clear, however, from recent problems that nuclear energy has significant unreliability.

Nuclear plants are generally constructed in capacity increments of 1000 MW$_e$. Since nuclear energy offers a utility a large increment of power for transmission, utilities have been more inclined to pay the enormously high corridor construction costs when a nuclear facility is connected to the transmission line. In the case of a corridor for geothermal energy alone, the risk and small amounts of energy (at least initially) are often not worth the reward.

There is a convenient solution to this dilemma. If a large nuclear or coal-fired plant is situated so its transmission lines run near or through an area of geothermal production, the geothermal electricity may be tied onto the nuclear/coal line. A relatively small amount of geothermal electricity "piggybacks" onto an economical nuclear/coal line. If geothermal capacity scales-up over a 30-yr period in the manner, say, of Figure 5-3, additional nuclear plants and power lines may also be added, allowing an increase in piggybacking. If additional nuclear plants are not added after, say, 15 years of geothermal capacity scale-up, a utility may be more willing to construct a line solely for geothermal transmission. The reason is that, after 15 years, the utility would have major geothermal capacity which would substantially reduce operating risks.

In the case of Imperial County, the first such nuclear-related corridor involved the proposed Sun Desert nuclear plant. This 2000-MW$_e$ plant was to have been owned and operated by San Diego Gas and Electric Company (SDG&E). Its proposed location, shown in Figure 8-3, was in southeast Riverside County just to the north of the the northeast corner of Imperial County. To route this electricity to a San Diego service area, SDG&E contracted a routing study with Wirth Associates of San Diego. The two routes recommended by Wirth to SDG&E are shown in the figure. While the Sun Desert nuclear plant proposal was under review by the State Energy Commission, geothermal transmis-

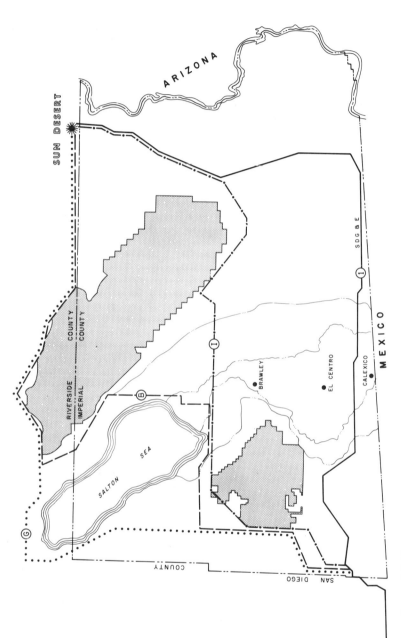

FIGURE 8-3. Transmission corridors from Sun Desert Nuclear Plant favored by a county selection committee in 1977. Routes B and G no longer apply, with the elimination of the Sun Desert plant. Routes 1 and I are similar to routes shown in Figure 8-1. (Source: Transmission Route Selection Committee, 1977)

Geothermal Development Update

sion routes were discussed in terms of piggybacking on some agreeable Sun Desert route.

In January, 1977, representatives of SDG&E and Imperial County public agencies, municipalities, and interest groups formed the Transmission Corridor Selection Committee (hereafter called Selection Committee), which met for nine months to decide on the best route alternatives through the county from Sun Desert to San Diego. In May, 1978, the State Energy Commission declined SDG&E's Sun Desert site permit request on environmental grounds, primarily because of nuclear waste disposal problems. From the present standpoint, it is unlikely that the Sun Desert nuclear plant will ever be built because of these problems. However, consideration is being given to locating a coal-fired plant at the Sun Desert site.

Many of the issues and conclusions of the Transmission Corridor Selection Committee are relevant to present corridor siting because of the recent development of a major site for nuclear energy in Arizona. A nuclear plant complex of 4000 MW_e capacity is under construction near Phoenix. Since its capacity could serve the energy needs of an area of about four million persons, Arizona cannot use all this nuclear energy. As a result, the California Public Utilities Commission has mandated to SDG&E the installation of a high-capacity 500-kV line to bring energy from Phoenix to the utility's San Diego service area.

High-capacity transmission routes under current consideration are shown in Figure 8-4. These routes were chosen as optimal in October, 1978 by the Utility Technical Committee of the Imperial Valley Action Plan. This committee, consisting of representatives of the major utilities, was appointed by the State Energy Commission to advise on corridor siting studies. The southern route, passing just above the southern border of the county, is the old SDG&E Sun Desert route— now intended to link with Arizona. The northern routes would carry energy from a fossil plant at Sun Desert or from the Arizona nuclear grid across the northern agricultural area of the county, up both sides of the Salton Sea to tie into the Los Angeles basin and across the West Mesa to tie into San Diego.

The earlier discussions and conclusions of the Transmission Corridor Selection Committee are valuable in bringing out the leadership dispute on corridor selection. The Selection Committee was composed of one representative from each of the following organizations: (1) San Diego Gas and Electric Company; (2) Imperial County Farm Bureau; (3) Imperial County Planning Commission; (4–10) Cities of Brawley, Calexico, Calipatria, El Centro, Holtville, Imperial, and Westmoreland; (11 and 12) Unincorporated areas of Ocotillo and Salton City; (3) Im-

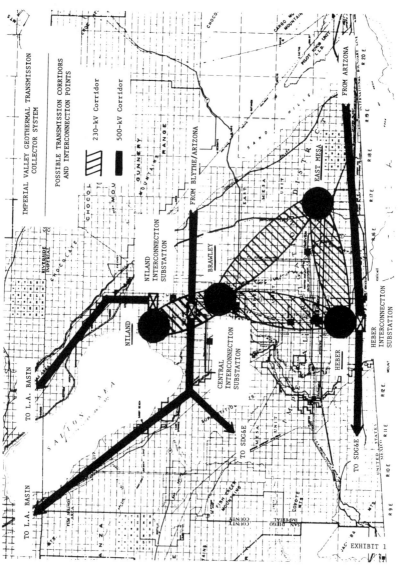

FIGURE 8-4. Routing recommendation of Imperial Valley Action Committee for a possible 200-MW$_e$ geothermal capacity. In this transmission corridor scenario, two very high voltage lines pass from nuclear plants in Arizona through the Imperial Valley (picking up geothermal power from the three substations shown) and on to consumption destinations in the San Diego and Los Angeles areas. The 230-kV lines serve as a collector system for geothermal power. (Source: Imperial Valley Action Plan, 1978)

perial Irrigation District; (14) American Association of University Women; and (15) National Parachute Test Range. The 1977 Selection Committee discussed issues and reached conclusions related to the present Arizona-to-California transmission requirements shown in Figure 8-4. In evaluating alternative corridor routes based on the different interests represented, the very wide range of potential impacts evoked different responses among participants. This broad range of potential impacts of transmission corridors is shown by the following list, compiled by Wirth Associates:

1. Geotechnical features
    a. faults
    b. earthquake epicenters
    c. soil conditions
    d. erosion potential
2. Ecological resources
    a. plants
    b. animals
    c. rare and endangered species
3. Cultural resources
    a. historical sites
    b. archeological resources
4. Land use
    a. agriculture
    b. urban areas
    c. park and natural areas
    d. highways
    e. utility rights of way
    f. general plans
    g. restricted areas
    h. socioeconomic impact
5. Scenic–visual resources
    a. unique landscapes
    b. vista points
    c. highway views

As the Committee meetings proceeded, each group presented its selection criteria and interests on a preferred route. In many cases, criteria and interests of several groups were in direct opposition. Several committee representatives, as well as some not on the committee, compiled lists of specific routing objections. Objections divided geographically into the East Mesa Section, Central Irrigated Valley Section, and West Mesa Section are shown in Table 8-1. The Bureau of

TABLE 8–1. AGENCY ROUTING OBJECTIONS FOR WEST, CENTRAL, AND EAST SECTIONS OF IMPERIAL COUNTY, AUGUST, 1977[a]

West section	Central section	East section
Bureau of Land Management		
Areas to avoid:	None	Areas to avoid:
1. Davies Valley		1. Sandhills north of Highway 78
2. South of Highway 98 in Ocotillo region		2. Any area closed to vehicular traffic
3. Close proximity to Ocotillo		3. Salt Creek Marsh
4. Coyote Mountains Primitive Area		
5. Fish Creek Mountains		
6. In-Ko-Pah Mountains Primitive Area		
7. San Sebastian Marsh		
8. Any areas closed to vehicular traffic		
Navy NPTR		
Areas of concern:	Areas to avoid:	Areas to avoid:
1. Area between landing field and Restricted area R-2510	1. Airport traffic area—comply with FAA Regulations	1. Chocolate Mountains Restricted Area R-2507
2. Restricted area R-2510 (NPTR test range)	2. Salton Sea restricted area R-2521	Area of concern:
		1. Restricted Area R-2512 (west of sand dunes)
State of California, Department of Parks and Recreation, Wildlife		
Areas to avoid:	Areas to avoid:	Areas to avoid:
Anza-Borrego State Park	1. Roadside rest stop—Seeley	Salton Sea State Park (Northeast shore of Salton Sea)
	2. Roadside rest stop—Calipatria	
	3. Salton Sea National Wildlife Preserve	
	4. Wister Waterfowl Management Area	
	Area of concern:	
	Finney Ramer Lake	
Imperial Irrigation District		
Area to avoid:	Areas to avoid:	Areas to avoid:
Parallel Encroachment of 92-kV line from Salton Sea to Imperial	1. Occupation of All American Canal right-of-way	1. Occupation of All American Canal right-of-way

2. Interference with operation, maintenance, or repairs of canals drainage and transmission systems
3. Encroachment of existing rights-of-way
4. Areas of agricultural usage (least interference possible)
5. Spheres of influence of cities

Farm Bureau

Areas of concern:
Agricultural land (least impact possible)

County of Imperial

Areas to avoid:
1. Previously subdivided areas not developed
2. Spheres of influence of towns such as Ocotillo, Plaster City, Salton City

Areas to avoid:
1. Diagonal crossing of agricultural field
2. Feedlots, agricultural processing plants, or houses
3. Parks—Sunbeam Lake, Wiest Lake, Heber Beach, Osborne Park
4. Airports (Comply with FAA Regulations)
5. Wildlife Areas
6. Spheres of influence of cities

Areas of concern:
1. Agricultural lands, Classes 1, 2, and 3, (least impact possible)
2. A-3 zones
3. Microwave line-of-sight transmission—Salton City, Black Mountain, Superstition Mountains, Brawley

2. Parallel encroachment of transmission corridors

[a]Most objections apply to corridor routing alternatives from the proposed Sun Desert Nuclear Plant shown in Figure 8-3. Most objections are pertinent to the present Arizona to San Diego routing dilemma and to present northern power routing dilemmas. (Source: Imperial County Selection Committee, 1977.)

Land Management manages most of the land on the East and West Mesas and expressed concern for avoidance of many ecological and scenic areas on the Mesas, but had no objections in the Central Section, while the State Department of Wildlife sought corridor avoidance of many state parks and areas. The concerns of the Naval Parachute Test Range (NPTR) were entirely different. NPTR sought preservation of security in areas on the Salton Sea and Mesas and maintenance of airport safety at its El Centro Naval Air Station.

Also outlined in the table are the concerns of three important local groups: the IID, the County Farm Bureau, and the County Planning Department. The IID voiced concerns about encroachment of existing power transmission lines on its irrigation canals, especially the All American Canal. These concerns were based on IID's desire to ensure expansion prerogatives for its existing distribution networks. The Farm Bureau was concerned about effects on farm land prices and on agricultural operations, such as on night spraying activities.

The objections of the County Planning Department echoed many concerns of the above groups—agricultural, park and recreational, airport, and ecological. In addition, the county stressed avoidance of cities. The latter objection was in response to the threat of the SDG&E primary route, which would run 1½ miles north of the city of Calexico. As Calexico is bounded to the south by the international border and to the east by the All American Canal, a northern delimitation would presumably cramp city growth. This issue becomes especially sensitive when it is recalled from Chapter 3 that 31% of the county's Spanish-American population is located in Calexico.

By the end of the Committee deliberations, 11 alternative routes had been proposed. Each representative on the Committee then voted on all 11 routes by giving a rating on a scale of 0 (full disapproval) to 3 (full approval). Table 8-2 shows the top four choices as determined by vote of the Selection Committee. The four routes are shown in Fig-

TABLE 8–2. THE FOUR MOST PREFERRED TRANSMISSION LINE ROUTES

Route	Agency proposing	Rating (out of 150)
G	City of Imperial	92.0
I	City of Holtville	65.0
B	American Association of University Women	65.0
1	SDG&E	62.2

FIGURE 8-5. Production field, Cerro Prieto.

ure 8-3. Route G, which was overwhelmingly preferred, loops widely to the north and west avoiding the agricultural central part of the county. Another similar route, Route B, goes widely north and west, but cuts south of the Salton Sea. Neither of these routes was recommended by the Imperial Action Plan Committee (Figure 8-4). For Route 1, there would be problems with establishing geothermal interconnections. For Route B, the extra length would be a problem since current construction costs of transmission corridors are about $800,000/mile. Routes I and 1, on the other hand, are closely duplicated in the Action Plan recommendations.

This discussion has emphasized a multifaceted political conflict already emerging on a critical geothermal issue. Another critical issue may evolve when the production field, wells, and pipes are in place, figure 8-5. As suggested in Chapters 6 and 7, there are many other latent leadership issues. The surfacing of these as real conflicts will depend on the sequence of geothermal development events. Such events may have an erratic and/or chance origin. Although still untested, the same conclusions would seem to apply to other latent public opinion issues.

Electrical Use Developments

Generation of electricity by geothermal power plants is a key economic step in geothermal development. It is the point at which exploration, production, field development, and transmission corridor construction come together to result in an economic exchange between producer and utility. Electrical use developments are discussed below and shown in Table 8–3 for the major KGRAs—East Mesa, Heber, Brawley, and Salton Sea.

East Mesa

The East Mesa geothermal field located in the southern part of the East Mesa and centered about 9 miles southeast of the town of Holtville, consists of 38,365 acres, of which 33,525 are federal land. It is a very large and relatively shallow reservoir with a reservoir top at a depth of 2,500 to 8,000 feet. Temperature is about 400°F, and salinity levels are 1,800–28,000 parts per million (ppm).

Two production companies are active on the East Mesa—Magma Power Company in the far south and Republic Geothermal to the north of Magma. As of February, 1977, nine wells were drilled, but each company has an ongoing field development drilling program. As may be observed in Figure 8-6, each company currently plans to build a 10-MW_e test plant, each to be followed in the early 1980s by standard-sized 50-MW_e plants. To aid in achieving these goals, in 1978, Republic Geothermal was issued $9 million under the Federal Loan Guarantee Program to drill 11 production wells and four reinjection wells, enough for a 35-MW_e plant. Republic has announced its power plant plans, and it is anticipated it will soon sign a contract with SDG&E to commence power plant construction.

In 1978, Magma Power Company began construction on an innovative 11-MW_e binary test plant. This is the first pilot plant greater than a megawatt in the valley which will produce electricity, since existing test facilities, the DOE facility on the East Mesa and the 10-MW_e SDG&E facility at Niland, do not have electrical generators. Magma's plant was designed by J. Hilbert Anderson, Inc., of York, Pennsylvania. The plant, completed in the Fall of 1979, produces power that is distributed on the IID collection grid to Imperial Valley customers.

The plant represents the world's most modern and technically advanced geothermal power facility, incorporating many features never before attempted. As seen from the schematic drawing in Figure 8-6,

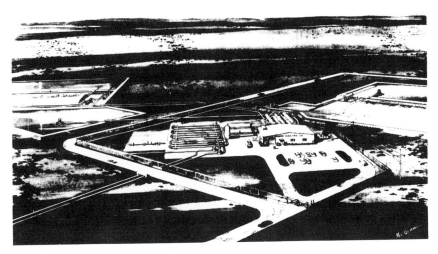

FIGURE 8-6. Artist's rendering of Magma Power's 11-MW$_e$ pilot plant in the East Mesa KGRA. Most plant equipment and piping are located outside. Cooling ponds are in the background. (Source: Dambley, 1978)

the plant is very spread out over 24 acres. This is possible in the East Mesa, because unlike other major KGRAS, no agricultural land is displaced. There are five production wells slant drilled to a depth of about 7000 feet. Slant drilling means that drilling takes place in one or several directions other than vertical. Three reinjection wells are located about one mile from the plant. Two of the production wells and one reinjection well will be kept out of use as spares.

The basic binary plant flow uses isobutane as the working fluid (Figure 2-10). However, there is a second binary loop, utilizing propane as working fluid, transferring heat from the isobutane turbine exudate of the first loop to lower-temperature propane in the second loop. Each loop is attached to its own turbine generator equipment—the first loop producing at a 10-MW$_e$ power capacity and the second loop at a 3-MW$_e$ capacity (these total MW$_e$ figures are higher than the net plant output of 11 MW$_e$, because some power produced is consumed by the plant itself). The York turbines are unique in using a patented Magmamax process. The engineering principle is that of standard refrigeration in reverse. In refrigeration, power is used to cause a temperature drop. In the plant, a temperature drop in the binary fluid is employed to produce power.

There are several other unique features:

Heat Exchangers

The ten exchangers are very elongated and designed to work using a true counterflow process rather than using conventional baffling. As a result, efficiency of heat exchange is increased and pumping power is greatly reduced.

Pumps

A standard problem in hot water production fields is that some wells will produce spontaneously, while others need stimulation. Also, spontaneous wells may show large temperature drops from well bottom to well top and scaling depositions in the bore holes. A pump employing concepts from Iceland's geothermal wells will reduce the above problems.

Cooling System

Since area is not a problem, three large cooling ponds covering 23 acres have been included in the design. The ponds are at different elevations, with two higher-level hot ponds for spraying and storage, and a lower pond for make-up cooling water. Cooling water circulates through the plant 24 hours a day. The hot ponds, which are lined with heavy vinyl plastic, have sprayers. To achieve more efficient cooling, spraying is done in the coolest 16 hours of the day, with storage during the other eight hours. Another unique environmental feature is the positioning of the plant to aim into the East Mesa's prevailing winds. The spray lines in each hot pond are at 23° angles to the prevailing winds—an angle offering maximal cooling. Brine discharge will be an important facet of all geothermal development in Imperial County, Figure 8–7.

Magma has estimated the final capital cost of the plant at $825/KW$_e$, a cost which will be seen later under Limiting Factor—Financing to be very reasonable for a binary pilot facility. Remarkably, the entire cost for this plant of over $9 million has been generated by the internal cash flow of Magma Power Company, a company with 1975 total assets of $26.5 million and net income $5.3 million.

After testing this facility, Magma has indicated tentative plans to follow with a 50-MW$_e$ East Mesa plant, presumably also binary. This innovative pilot plant represents the initial and lead plant for geothermal development in the Imperial Valley.

FIGURE 8-7. Brine discharged into evaporation pond, Cerro Prieto.

Heber

The Heber geothermal field (see Figure 1–2) is centered about six miles southeast of El Centro. The KGRA consists of 58,568 acres of land, all privately owned. The top of the reservoir is at 3200 feet in depth. Temperatures range from 350 to 375°F and salinity is about 14,000 ppm.

Although both Chevron Resources and Union Oil Companies have land holdings and drilling experience in Heber, Chevron has the larger lease holdings and is most likely to be the sole field operator, with Union remaining a passive minority participant. This situation is just the reverse at the Brawley field, where Union Oil is the operator. Up to 1978, Chevron had drilled nine test wells and Union had drilled seven test wells.

The history of power plant prospects at Heber has centered around an attempt at a federal-funded pilot project. The pilot idea stemmed from a study by the Electric Power Research Institute (EPRI) in Palo Alto, indicating Heber would be a preferred location for a full-size federally funded demonstration plant. EPRI subsequently committed over $2 million in funding for preliminary environmental and engineering studies of this site. Meanwhile, SDG&E submitted a proposal to DOE for partial federal funding of a 50-MW$_e$ demonstration binary plant on the Heber site, with Chevron as producer. "Demonstration" implies that other utilities, engineering companies, and agencies will have full access to the testing and production data at the plant.

In mid-1978, however, the Heber demonstration project proposal was rejected by DOE and the funding was awarded instead to a 50-MW_e demonstration plant at the Valles Calderas field in New Mexico. Even though considerable study went into the Heber pilot project plans, it appears to have been eliminated from consideration for the present. Instead, initial development at Heber will be a 50-MW_e two-stage flash plant to be built with private sector funds by SCE in conjunction with Chevron as producer. A contract has been signed, and a plant will likely be operating by 1984. Chevron plans to drill nine production wells and six reinjection wells. Make-up cooling water will come from the New River. Another planned feature of this plant is sophisticated field engineering borrowed from oil industry expertise, to accurately "mine" the heat in the field. Subsequent Chevron/SCE or Chevron/SDG&E plants at Heber will come in line sequentially up to a total of perhaps 400 to 500 MW_e.†

Brawley

The Brawley field, centered about three miles north of the town of Brawley (Figure 1–2), lies underneath a KGRA consisting of 28,885 acres of private land. It is a very hot field (about 500°F) and highly saline (100,000 ppm). The reservoir top is at a depth of 3,000 feet.

The Brawley development is dominated by Union Oil and SCE. Although Chevron owns a portion of the land leases and has drilled two wells, Chevron probably will have only a nonoperating minority interest at Brawley. By fall, 1978, Union Oil had drilled six wells and was in the process of drilling and permitting 14 additional ones.

Union's initial power plant plan is to construct a 10-MW_e single flash pilot plant to go on line in the early to mid-1980s. The plant engineers will be Rogers Engineering of San Francisco. The major problem to be surmounted by pilot plant testing is scaling and corrosion from the high-salinity brines. Because of the salinity, it is likely that the geothermal condensate will be reinjected and fresh canal water will be used as make-up water for the first full-size plant. Beyond the test plant, a contract has been signed by Union and SCE for development of the field up to a maximum of 450 MW_e. The Brawley field is estimated to be very large with some estimates exceeding 1000 MW_e.

†More recent information suggests that a binary demonstration plant may be built in the Heber area. Also, now there are two 10-MW_e plants operating in Imperial Valley.

TABLE 8–3. ESTIMATED SCHEDULE OF ON-LINE POWER PLANT ADDITIONS: IMPERIAL COUNTY

			Plants on-line by year:								
Area	Utility developer	Size (MW$_e$)	1979	1980	1981	1982	1983	1984	1985	1986	1987
Salton Sea	Magma/NARCO	50				X					
	SCE/Mono-Union-South Pacific	50				X					
	SCE	50									X
Westmoreland	SDGE/RGI-MAPCO	48					X				
East Salton Sea	DWR/McCulloch	55						X			
Brawley	SCE/Union	10		X							
	SCE	100						X			
	SCE	100							X		
	SDGE/Chevron	100								X	
	SCE/Chevron	100								X	
South Brawley	DWR/CUI Venture	55						X			
East Mesa	SDGE/RGI	48				X					
	SDGE/Magma	10	X								
	SDGE/Magma	40[b]						X			
Heber	SCE/Chevron	50				X					
Cumulative power on-line:			10	20		218	266	516	616	816	866

[a] Source: Science Applications, 1978 and California Energy Commission, 1979.
[b] Additional capacity—expansion of 10-MW$_e$ plant.

Salton Sea

The Salton Sea geothermal field is the largest and hottest of the Imperial County fields, and one of the largest worldwide. It has a KGRA area of 95,824 acres, 18,644 of which are federal. The brine temperature, in places, exceeds 640°F, and the depth of the top of the reservoir is 3,000 feet. In otherwise favorable circumstances, a problem is an exceedingly high salinity of 250,000 to 330,000 ppm (by comparison, the salinity of sea water is 35,000 ppm).

Since the deep reservoir was discovered in 1958, over 25 wells have been drilled, but a number of them have been abandoned. A 10-MW_e nongenerating pilot test facility was constructed in 1975 by SDG&E with ERDA (predecessor to DOE) funding. This has been mostly used to yield test results related to high-salinity brines. Currently, the leading developers of this field are Union Oil and Magma. Union Oil has entered into an agreement with Southern Pacific Land company and the Mono Power Subsidiary of SCE to develop the Salton Sea reservoir. The consortium has rights to the KGRA's largest combined landholding, and Union has a 50% right to the group's lands. As initial steps, Union will conduct several years of fluid tests, and SCE will construct a 10-MW_e pilot plant in the early 1980s. Because of technological unknowns, it is difficult to predict the consortium's pace of development thereafter.

FIGURE 8-8. Alfalfa plant, El Centro.

FIGURE 8-9. Valley Nitrogen plant, El Centro.

Magma Power, in conjunction with SDG&E, has recently announced completion of a contract for construction by Morrison Knudson Co. of Boise, Idaho, of a 50-MW$_e$ plant at Niland. This plant will use flash technology with a "double" entry turbine. Magma's project, however, is conditional on completion of the regulatory process, which might take several years. Another proposed project in the general Salton Sea area is a Republic Geothermal plan for a 50-MW$_e$ plant in the vicinity of the town of Niland.

DIRECT USE DEVELOPMENTS

Use of geothermal energy is not restricted to generation of electricity. Imperial County is presently anticipating use of the geothermal hot water in two projects largely sponsored by the Department of Energy. One is a project to geothermally heat and air condition the City Community Center in El Centro and a second project located at the Holly Sugar plant between El Centro and Brawley is to geothermally dry sugar beet pulp. An alfalfa plant in El Centro is shown in Figure 8–8. Another potential project, presently unfunded but nevertheless carefully studied, is to use geothermal turbines at the Valley Nitrogen fertilizer plant located adjacent to Heber, Figure 8–9.

In the present section, these three project plans will be discussed in detail. Prior to this discussion, however, some general background on direct use is given. Direct use means applying the heat of geothermal brines to the heating and cooling of residences, industrial processes, and agriculture. The minimum temperature of geothermal brines for direct use may be considerably lower than that for electrical use. Resource temperatures as low as 60°F may be used for heating and cooling of small buildings. The lowering of temperature limits means that a much larger resource base exists for direct use than for electrical. Lienau (1978) has estimated that low- and medium-temperature resources (194–300°F) are four times as abundant as those of high temperature. Since various direct use applications utilize temperatures ranging from 60°F upwards to 400°F, some project designs make use of the principle of *cascading*. This means that a high-temperature brine is used for an initial direct use application, leaving a residual brine flow of lower temperature. This residual flow may then be utilized for a different medium-temperature application, leaving in turn another low-temperature residual flow, and so on. For such uses, geothermally heated water can be transported in pipes up to a distance of about 500 miles from the well site.

The variety of industrial direct use applications are shown in Table 8–4. Such uses date back early in geothermal history. For example, houses have been geothermally heated in Iceland for over 50 years. Worldwide, geothermal energy being utilized in direct use far exceeds that for electrical use. Schultz et al. (1978) estimated that currently an equivalent energy capacity of 5,500 megawatts (thermal) exists for direct uses. This compares to 1976 worldwide geothermal electrical capacity of 1,362.2 megawatts (electrical). For direct uses in the United States, however, an equivalent capacity of only 50 megawatts (thermal) is presently used.

Present direct use capacities merely scratch the surface of potential future capacity. For district heating of houses and buildings in the 11 western states, Lienau (1978) has estimated a district heat demand, which could be provided by geothermal means, of 478.8×10^{12} BTU/year, an equivalent capacity of 15,942 MW_t. His reasoning is demonstrated in Table 8–5. He calculated that 15% of the U.S. population resides within 50 miles of the 224 water-dominated geothermal areas. He estimated such a population's heating demand in direct geothermal use would account for 3.4% of the country's present energy consumption for space heating.

For U.S. industry, the potential for geothermal applications is also large. In 1975, 20.4×10^{15} BTU of energy was consumed by U.S. industry. Schultz cites the categories for industrial end use of this energy in Table 8–6. Geothermal water-dominated resources could potentially supply energy for the first two categories, accounting for 68.4% of U.S. industrial uses. Although geography of plant locations and economics

TABLE 8-4. INDUSTRIAL PROCESSES UTILIZING GEOTHERMAL ENERGY[a]

Industrial process	Country	Description of industrial process
Wood and paper industry		
Pulp and paper	New Zealand	Processing, plus small electrical power production
Timber drying	New Zealand	Kiln operation
Wood washing/drying	Iceland	Steam drying
Mining		
Diatomaceous earth plant	Iceland	Dried earth recovered by wet mining techniques
Chemicals		
Salt plant	Japan, Phillipines	Salt from sea water
Sulfur mining	Japan	Sulfur extraction from volcano gasses
Boric acid ammonium bicarbonate and sulfate, sulfur	Italy	Recovery from noncondensable gasses in steam
Miscellaneous		
Confectionary industry	Japan	
Grain drying	Phillipines	Steam heats kiln dryer
Brewing/distillation	Japan	
Stock fish drying	Iceland	Fish drying in shelf dryers
Curing cement slabs	Iceland	Light aggregate cement building slabs
Seaweed	Iceland	Drying seaweed for export
Onion drying	U.S.	Dehydration of onions
Milk pasteurization	U.S.	Uses low-temperature resource

[a] Source: Schultz et al. (1978).

considerably reduces this potential, what remains is, nevertheless, enormous relative to present use.

From the above discussion, a problem in direct use is location of residences and industries within 40-50 miles of geothermal hot water areas. Two other problems loom large in realizing the great potential for direct uses. One of these problems is the economics of wells and transmission pipes for direct use. A second problem is the economic practicality of adapting the engineering design of existing residences, industrial plants, etc., to geothermal direct use purposes (the retrofit problem).

Regarding the economic problem, it is important to realize that although technical practicality of wells and pipes may be present for a particular direct use project, economics can often, nevertheless, rule

TABLE 8-5. POPULATION SURROUNDING WATER-DOMINATED GEOTHERMAL AREAS IN THE ELEVEN WESTERN STATES[a]

Radial distance from the resource (miles)	Total population within radius	District heat demand[b] (10^{12} Btu/yr)	Megawatt (thermal) equivalent[c]	Billion barrels of oil equivalent[d]
0–4.99	120,000	1.9	63.3	0.0004
5–9.99	330,000	5.7	189.8	0.0013
10–19.99	3,950,000	74.9	2,493.8	0.0174
20–29.99	6,070,000	86.3	2,873.4	0.0200
30–39.99	14,210,000	180.1	5,996.5	0.0418
40–49.99	9,670,000	129.9	4,325.1	0.0301
Totals:	34,350,000	478.8	15,941.9	0.1100

[a] Source: Lienau (1978).
[b] Total U.S. district heat demand = $13,900 \times 10^{12}$ BTU/yr.
[c] Total U.S. megawattage (thermal) equivalent for space heating = 463,000 MW.
[d] Total U.S. oil consumption = billion barrels/yr.

the project out. This topic is addressed in detail later in the chapter under "Limiting Factor—Transmission Pipes."

Solution to the retrofit problem of adapting existing structures and plants to direct uses depends on technology, economics, and the advantages offered by a particular water-dominated resource. Geothermal house heating technology is readily adaptable to U.S. residences and buildings that are space conditioned by four types of systems: forced air, convectors, radiant panels, and heat pumps. In industrial plants, new and often experimental designs will be required for adaptation. Some modified or newly designed geothermal equipment will be required, at least in trial cases. Examples of plant equipment adaptation are discussed later for the Holly Sugar and Valley Nitrogen plants.

City of El Centro Community Center†

The first geothermal space conditioning project in Imperial County will involve the drilling of a geothermal well within the El Centro city limits and the transport of geothermally heated hot water several blocks in order to heat and cool a 13,200-ft² city Community

†These are the major direct use projects; other direct use projects are now underway in Imperial Valley.

Center building. The Center is used for a variety of recreational and educational community activities. This $2.7 million project is 72% funded by the U.S. Department of Energy, with the remaining funding by the City of El Centro.

The project participants are the City of El Centro, Chevron, Inc., and Westec Services, Inc., a geothermal contractor located in San Diego. The city owns the Community Center building and property, the drill site, and pipe transport lands, as well as the right to the geothermal resources underneath the drill site and the right to reinject the spent brines after use for heating/cooling. After the end of the experimental Community Center project in several years, the city will retain ownership of all land and facilities involved. Chevron, Inc., mentioned in the last section as a major geothermal electrical-use developer in the valley, will be responsible for the drilling and resource production. Westec Services Inc., will be the city's contractor for the nondrilling stages of the project, including construction of transmission pipelines and the heating/cooling system for the center. The project timetable calls for drilling and production engineering in 1979, and installation of the distribution and heating/cooling system in 1980.

Only about 10% of the well flow will be utilized by the Community Center. For the remaining 90% of the well flow, the city has further plans to develop under HUD sponsorship an industrial park nearby for a variety of geothermal projects. Beyond the full use of the single well flow, the city envisions the formation of a geothermal municipal utility to distribute geothermal heat in El Centro for various types of end uses.

The center is a heavily used multipurpose facility. It was built in 1974 at a cost of $520,000 under a HUD grant and is located on 2.7

TABLE 8–6. CATEGORIES OF ENERGY USE BY INDUSTRY

Use	Percent
Process steam	40.6
Direct process heat	27.8
Electric drive	19.2
Feed stocks and chemicals	8.8
Electrolytic process	2.8
Other	0.8
Total:	100.0

acres in the southeast part of El Centro, a part of the town shown in Chapter 3 to have low income and high percentage of ethnic minorities. The Center has activities and social events for up to 1,500 persons, English and other classes, daycare for infants, meetings of outside organizations, a small city branch library, crafts, games, and sports. The Center staff of seven accommodates an average of 700 persons per day, remaining open seven days a week. Presently, the building is adequately heated and cooled by natural gas.

The availability of geothermal energy to the city of El Centro was shown on the KGRA map in Figure 1-1. The boundary of the Heber KGRA runs just south of the city's southeast corner. The town of Heber, 4½ miles South of El Centro, has the KGRA's maximum temperature resource of about 375°F. At the 2.5 acre drill site near the Community Center, a geothermal gradient is expected of 2.0 to 3.3°F. per 100 ft. The drilling proposed by Chevron for this site consists of a 7,000-ft production well of 8⅝-in. diameter, and a 4,000-ft reinjection well. In addition, the project plans contain an option to drill an even deeper 10,000-ft well, to be used for geological borings and temperature gradient studies.

A distribution system will run the three blocks from the drill site to the Center building. This system will employ a surface heat exchanger, in order to eliminate the transport of hot brines along the city streets. Figure 8-10 illustrates the distribution system of this exchange mechanism. Brines are brought to the surface by the production well, flowed through the heat exchanger located at the drill site, and returned by reinjection to the geothermal reservoir. From the heat exchanger, a separate close loop of clean hot water flows along the 8-in.-diameter transport pipes to the Center, where it heats space heating coils to perform the heating/cooling function. It is also used as a source for domestic hot water. The clean hot water is then returned via transport pipes to the heat exchanger, to start the hot water loop again.

A separate chilled water loop originates with a lithium bromide absorption unit also located at the drill site and receiving part of the hot water loop. The engineering principle followed here is that of commonplace refrigeration. Water chilled by the absorption unit flows along a separate set of 10-in.-diameter transport pipes to the Center, cools the building by means of space cooling coils, and returns along the pipes to the absorption unit. Location of the heat exchanger/refrigeration equipment at the drill site offers the potential for future expansion into a true district heating arrangement, in which many pairs of hot and cold water loops would originate in the exchangers at the drill site. The Community Center's geothermal heat-

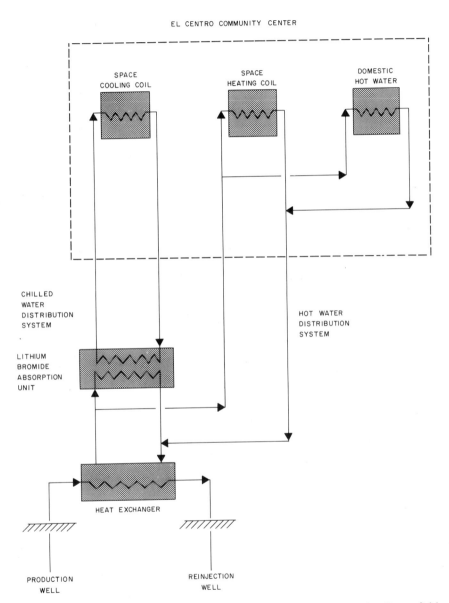

FIGURE 8-10. Proposed water distribution system for El Centro Community Center field experiment. Two clean water loops—one hot and one cold—distribute the energy in the geothermal brines from the well site ⅓ mile to the Community Center. (Source: City of El Centro, 1978)

TABLE 8-7. THE COMMUNITY CENTER PROJECT BUDGET[a]

Budget item	City share	DOE share	Total cost	Percent
Geothermal heat supply	23	1170	1193	45.0
Cooling/heating system	615	479	1094	41.2
Start-up/evaluation	20	187	207	7.8
Administration	79	54	133	5.0
Public relations	2	24	27	1.0
Totals:	739	1914	2654	100.0

[a] In thousands of 1978 dollars.

ing/cooling system consists of the heating and cooling coils and rooftop air handling units to serve zones of the building. Temperature and pressure control for the hot and cold air circulation loops is achieved by a sophisticated system of electronic gauges.

The direct use application just described is quite costly, but may be justified in its cost by providing the pilot project and nucleus of a large, municipal, direct use distribution system. The budget for the Community Center project is given in Table 8-7.

This budget may be compared to that cited by Lienau for a $18.9 million district space heating system in Akureyri, Iceland. For Akureyri, geothermal heat supply accounted for only 24% of costs, with the heating/distribution system accounting for the remaining 76%. The discrepancy in the two budgets is due to the extreme depth of the El Centro well and the mere 10% distribution system in El Centro. In fact, in El Centro, there is no possibility of paying back the $2.6 million capital costs, unless the entire well flow were to be utilized.

The true value of the Community Center project does not lie in good return on investment, but rather in stimulation of other direct use applications in the city. The city's energy supply presently comes from electricity (6.13×10^{11} BTU/yr). The end uses of these energy flows are presented in Table 8-8. It is not surprising in a locality of such extremely hot summer temperatures that air conditioning accounts for nearly 40% of energy use. Air conditioning and the other energy consumption categories potentially convertible from present energy sources to a geothermal source account for 69% of energy consumption. Thus, the potential exists to convert most of the city's energy needs to geothermal. As seen in Table 8-9, however, conversion to geothermal may decrease the efficiency of utilization from the level of current energy sources. However, this analysis ignores the electricity used in generating the current oil energy source.

The key question here, however, is whether conversion of a large portion of the city to geothermal energy would stand on its own as truly cost effective, without federal outlays. One benefit of the Community Center project is that such an economic cost/benefit analysis for the whole city will be possible, based on capital and operating cost data collected in the project. Another potential benefit is the likely advent of pilot projects on direct uses in El Centro which are new and unrelated to the existing uses given in Table 8–8. If the city succeeds with plans for a HUD-sponsored industrial park development adjacent to the Community Center drill site, industry might be encouraged to locate in the park and experiment on new uses. Several potential new uses proposed by the city are listed in Table 8–10.

Holly Sugar

The Holly Sugar plant located between Brawley and El Centro is one of the largest sugar beet processing facilities in the U.S., producing about a million pounds of high-purity sugar per day (TRW, 1977). Funding has been given by the Department of Energy to Holly Sugar and TRW Inc. in the amount of $10 million to redesign the plant for 70% dependence on geothermal energy and to drill a production and

TABLE 8–8. Energy Consumption by End Use for City of El Centro.[a]

End use	BTU/yr ($\times 10^{11}$)	Percent
Air conditioning[b]	3.7	37
Water heating[b]	1.5	16
Space heating[b]	1.1	11
Refrigeration		
Residential	0.6	6
Commercial[b]	0.5	5
Cooking	1.0	10
Residential clothes drying	0.2	2
Other	1.2	13
Totals:	9.8	100

[a] The yearly end use energy total is supplied by 179.6 million kilowatt hours of electricity and 369.5 million cubic feet of natural gas. 69% of uses are potentially replaceable by geothermal energy. Source: City of El Centro (1978).
[b] Potentially replaceable by geothermal energy.

TABLE 8-9. GEOTHERMAL HEAT REQUIRED FOR MAXIMUM GEOTHERMAL USE IN THE CITY OF EL CENTRO[a]

End use	Energy consumed from present sources (10^{11} Btu/yr)		Utilization efficiency of present energy sources		Actual heating/cooling load (10^{11} Btu/yr)		Geothermal utilization efficiency		Geothermal heat required (10^{11} Btu/yr)
Air conditioning	3.7	×	2.0	=	7.4	=	0.55	×	13.45
Water heating	1.5	×	0.64	=	1.0	=	0.92	×	1.09
Space heating	1.1	×	0.75	=	0.8	=	0.95	×	0.84
Commercial refrigeration	0.5	×	2.0	=	1.0	=	0.55	×	1.82
Totals:	6.8				10.2				17.20

[a] Conversion factors are given from present energy sources to a heating/cooling load and from geothermal heat to a heating/cooling load. On the average, geothermal energy requires 2.5 times as much energy. Source: City of El Centro (1978).

TABLE 8–10. POTENTIAL INDUSTRIAL APPLICATIONS OF GEOTHERMAL ENERGY FOR THE CITY OF EL CENTRO[a]

Potential uses
Agriculture
Lactose production
Seed processing
Vegetable, fruit, and floral greenhouses
Livestock feed processing
Asphalt plant
Furniture manufacturing
Vegetable and fruit packing operations
Vegetable, fruit, and other processors (frozen, canned, and dried)
Powdered milk and dairy product plants
Spa and health resorts
Refrigerated storage plants
Ice plants
Agricultural chemical production
Commercial laundry
Molded paper, cardboard, and plasterboard production

[a]Source: City of El Centro (1978).

reinjection well. The full $18 million project entails the building, operation, and evaluation of the redesigned plant. Results are transferable, because several of the 55 U.S. sugar beet factories are located within transportable distances of water-dominated geothermal resources of temperature greater than 300°F.

The plant location is not within a KGRA boundary, but lies between the Brawley and Heber KGRAs (see Figure 1–1). Nevertheless, the temperature of the resource near the plant has been estimated as 300–350°F. The total project is planned to take place in four phases over about a five-year period. Phase I, currently funded, encompasses drilling of a production and reinjection well, design of a pilot plant, and building of a pilot drier and boiler based on geothermal heat sources. In phase II, pilot implementation will take place at the plant with four production wells, three reinjection wells, 13 geothermal-designed driers and six geothermal-designed boilers. In phase III, the plant will be converted fully to geothermal operation. In phase IV, add-on geothermal capabilities will supplement geothermal drying and steam boiling of earlier phases. These capabilities include geothermal power generation and refrigeration.

Heat transport will differ from that in the Community Center project. At Holly, geothermal brines will be transported to the plant, run through a heat exchanger, and be transported back to the production area for reinjection. With such a brine transport loop, the danger of scaling of the transmission pipes and heat exchanger remains a critical question.

What are the economics and ultimate advantages of this project? Most of the total project cost ($10 million) lies in construction of the boilers and driers. Clearly, a federal subsidy is necessary to develop the initial new technology for these parts. Once developed, the equipment should be cost effective for future uses. Presently, the plant uses 351,000 barrels of oil per year. With geothermal retrofit on the present plant, this would be decreased to 122,000 barrels per year, a 65% reduction. For a new plant, only 45,000 barrels per year would be used. The potential of large future rises in the price of oil makes geothermal conversion a very attractive alternative. In addition, cost effectiveness may be increased if brine utilization is increased by a variety of non-sugar-beet uses. The sugar beet harvest season runs from April to August, but processing of sugar may be extended for 5 to 6 additional months by storage of concentrated juice. Geothermal energy would be mainly used during the harvest, leaving a partially or fully unused geothermal flow for half the year. Studies have shown potential offseason uses of this brine flow for gasahol production, alfalfa drying, and other uses. Hence, a project design which has the potential economic independence should emphasize multipurpose year-round use.

Valley Nitrogen

The Valley Nitrogen, Inc. plant is a major producer of nitrogen fertilizer. It is located south of El Centro near the town of Heber, the center of the Heber KGRA. The plant is owned by an agricultural cooperative of growers. In accord with the national trend of increases in the energy intensity of agriculture (especially in fertilizers), the plant presently receives the large annual energy supply of 9.1 billion cubic feet of natural gas and 66 million kilowatt hours of electricity. The latter is equivalent to the annual electrical generation of a 7.5-MW_e power plant having no capacity losses.

Although the fertilizer plant is not presently funded by the Department of Energy for conversion to geothermal energy, it is probable that geothermal conversion will take place at some point. Such an expectation is based on the plant's excellent location near geothermal fields and on a favorable economic cost/benefit picture for conversion

Geothermal Development Update

to geothermal power. The latter conclusion was drawn by Sherwood (1978) of Westec Services, Inc. in a recent economic analysis.

Although the following discussion applies only to the Valley Nitrogen facility, it illustrates the types of engineering–economic considerations involved in potential conversion of plants for other types of geothermal applications, such as those indicated in Tables 8–3 and 8–10.

The current plant design has three major steam systems at pressures of 50, 450, and 1500 psi. These systems receive steam produced from fossil fuel and from waste heat. The steam is used by 20 full-condensing, extraction, and back-pressure turbines, all potentially replaceable by geothermal turbines. The turbines are used at various steps in fertilizer production. Turbine energy requirements are concentrated in the three largest turbines: a synthetic gas compressor (power rating of 14,000 hp), a refrigeration compressor (7,325 hp), and a process air compressor (5,625 hp). The combined horsepower rating on the remaining 17 turbines is 6,696 hp, so that the largest three consume about 80% of the plant's fossil fuel.

There are many alternative ways to convert this plant to geothermal energy. First, any number and combination of turbines in the steam system may be replaced by geothermal turbines. Table 8–11 presents economic results of various turbine conversions for both flash and binary turbine technology.

To compute the internal rate of return shown in the table for conversions, the capital investment in geothermal conversion is subtracted from the energy cost saving of geothermal fuel over fossil. The annual energy cost saving is computed by the following formula:

$$S = C\,(a/f - b/g)$$

where

S = annual cost savings (dollars/yr)
C = annual net power output (Btu/yr for a geothermal powered plant)
f = net thermal efficiency of a fossil-powered plant
g = net thermal efficiency of a geothermal-powered plant
a = cost of fossil fuel (6 dollars/10 Btu)
b = cost of geothermal fuel (6 dollars/10 Btu),

with C calculated as follows:

C = net power (kW) × 3412.2 Btu/kWh × 8760 hr/yr × capacity factor.

The cost savings formula assumes that the net power output C is the same for a geothermal-configured plant as for a fossil-configured plant.

TABLE 8–11. ESTIMATED COSTS, BENEFITS, AND INTERNAL RATES OF RETURN FROM CONVERSION TO GEOTHERMAL POWER OF VALLEY NITROGEN PLANT, EL CENTRO[a]

Plant size (MW)	Number of turbines replaced	Flash				Binary			
		Capital investment in geothermal conversion[b]	Energy cost savings[c]	Incremental benefit-cost ratio[d]	IRR[e]— 10 years	Capital investment in geothermal conversion[b]	Energy cost savings[c]	Incremental benefit-cost ratio[d]	IRR[e]— 10 years
5	1	9.4	6.8	0.72	3.9	6.2	5.9	0.95	9.1
10	2	13.6	14.1	1.73	10.7	10.5	12.0	1.42	12.7
15	3	18.7	21.2	1.39	12.6	14.1	18.1	1.69	15.4
20	3	22.7	28.5	1.82	14.8	18.5	24.1	1.36	15.7
25	4	25.4	35.8	2.71	17.6	21.6	30.3	2.00	17.3
30	12	33.6	42.8	0.85	15.2	27.6	36.2	0.98	15.8

[a] Internal rates of return greater than 10% are considered worthwhile capital investments by management. Source: Sherwood (1978).
[b] In millions of dollars
[c] Present worth of energy cost savings over first 10 years (at 10% interest) in millions of dollars.
[d] See text for explanation.
[e] IRR = internal rate of return

Geothermal Development Update

The table reveals high internal rates for nearly all conversion possibilities. Only in the case of replacement of a single turbine are rates of return lower than the 10% return rate considered minimal, according to Sherwood, by the Valley Nitrogen management. Rates are comparable for flash or binary turbine technology, with binary slightly favored. They are most favorable for replacement of four turbines. The incremental cost benefit ratio is the ratio of the incremental energy cost savings to incremental capital investment in progressing from one turbine category to the next. It indicates a peaking of benefits at the category for replacement of four turbines.

The study also estimated geothermal and environmental effects of alternative plant designs. In Table 8-12, these effects are presented for a case in which only the three largest turbines are replaced by geothermal steam turbines. In this table the term "incremental cooling water" refers to the amount of cooling water needed above the amount used in present plant design (similarly, for the term "incremental canal water"). The hybrid alternative shown refers to replacement of the largest turbine by a binary two-stage turbo-expander with the next two largest turbines replaced by geothermal flash steam turbines. Under two of the binary alternatives shown, generation of part or all of the plant's electrical power would be produced by geothermal power generators installed in the plant.

The complication of geothermal modifications is apparent in Table 8-12. For instance, although the binary design with no on-site power generation has the highest internal rate of return (34.6%) and the lowest brine flow, it has twice the incremental water demands of the hybrid design. Although the binary alternative with fully on-site power has the advantage of avoiding any dependence on future utility sales of fossil-created electricity, it has a very high plant modification cost ($34 million) and a low internal rate of return (7.3%). These comparisons ignore tax advantages favoring conversion, such as the 20% tax credit.

Of the conclusions drawn from the Sherwood study, the following are the most important:

(1) For existing back-pressure turbines (in the smaller turbine range), it is more economical to use fossil fuel than to convert to geothermal energy.

(2) It is most economical on electrical generation to remain with off-site electricity supplied by the IID. However, obtaining future power blocks of 20 MW (the requirement given in Table 8-12) may be quite uncertain.

TABLE 8–12. DETAILED ECONOMIC SUMMARY OF ALTERNATIVE GEOTHERMAL MODIFICATIONS TO THE VALLEY NITROGEN PLANT, EL CENTRO[a]

Parameters	Hybrid: No on-site power generation	Flash: No on-site power generation	Binary		
			No on-site power generation	Some on-site power generation	Fully on-site power generation
Brine flow (10³ lb/hr)	3,237	3,701	1,354	3,084	4,155
Total plant power demand (kW)[b]	19,600	18,100	20,900	24,200	26,200
Offsite power demand (kW)	19,600	18,100	20,900	8,000	0
Onsite generator output (kW)	0	0	0	16,200	26,200
Incremental cooling water (gpm[c])	7,000	14,000	14,500	51,200	73,800
Incremental canal water (gpm[c])	−120	175	200	1,060	1,580
Plant modification cost[d]	21,927	22,122	15,153	26,715	34,424
IRR[e]—10 years	18.4	17.1	34.6	14.5	7.3
Incremental benefit–cost ratio[f]	0.86	0.68	2.91	2.18	0.64

[a] All five analyses assume that the three major compressors will be driven by geothermal turbines. Reservoir temperature is assumed to be 360° F. Total compression work is 24,532 hp. Source: Sherwood (1978).
[b] Includes present 8000-kW demand.
[c] Gallons per minute.
[d] In thousands of 1977 dollars.
[e] IRR = internal rate of return.
[f] See text for explanation.

(3) About 80% of possible cost savings comes from redesign for geothermal energy of the three largest turbines.

(4) Binary turbine technology is favored over flash. The principal reason is that capital costs are higher for flash conversion due to higher turbine costs and higher costs for transport of two-phase brine.

(5) For all modifications, large additional amounts of cooling water are required.

(6) With the largest three turbines converted to geothermal energy, the internal rate of return is higher than the minimal 10% required by management for all cases except that of binary with fully on-site electrical generation. Thus, some continuing electrical purchases appear essential.

This section has examined direct use development in the county. In the three instances discussed, some equipment involves new technology undergoing experimentation. These examples involve possible capital outlays ranging from $2.6 million to $35 million. Such capital outlays far exceed those of many existing instances of direct use, such as those in Boise, Idaho and Klamath Falls, Oregon. The reason for higher costs in Imperial County is mainly high drilling costs stemming from the much greater depth of the resource in the county. Nevertheless, there are unique opportunities, given carefully planned federal support, to convert many large existing plants and facilities in the Imperial Valley to direct use.

Regulatory Permitting Process in the County

Any geothermal development in Imperial County has to proceed through a regulatory permitting process. Although many current projects are presently at different phases in this process, only several are fully permitted and ready to proceed with operations. At the time of this writing (January, 1979), three 10-MW_e pilot power plants in Niland (SDG&E), and the East Mesa (Republic and Magma) are fully permitted and the 50-MW_e plant in Heber (Chevron Resources) and the net 48-MW_e Republic plant at East Mesa are fully permitted. In the regulatory process, type of land ownership (private, state, or federal) is of critical importance. Exploratory and development permitting for the three types of land ownerships will be discussed in order.

For private and state lands, the entire regulatory process generally takes about eight years (Lindsay, 1978). Two major steps in the process are embodied into laws. First, by a 1965 law, the State Division of Oil and Gas is the legal regulatory agency for well drilling and well operations. Second, by a 1970 California law, an Environmental Impact Statement (EIS) is required for any development potentially harmful to the environment. Beyond these legal requirements, a lead agency is always established for a particular project. A lead agency is defined as that federal, state, or local agency having the responsibility for assembling all pertinent information on a geothermal project and for approving the overall environmental plan.

In the case of private lands, the starting point of regulation is the negotiation of a lease between a private landowner and a developer for exploration and development. The lead agency, up to the end of the drilling process, is the county containing the resource. The county performs its regulatory function through local zoning regulations. Well drilling is regulated by the State Department of Oil and Gas. EISs are prepared by the county, but state agencies must approve them. Other regulatory bodies such as the State Department of Fish and Game and the local Regional Water Quality Board give their approvals on specific areas of interest. In the instance of state-owned lands, the lead agency, through the end of drilling, is the State Lands Commission.

Geothermal exploration and development on federal lands is based on a different legal framework from that of private and state lands. The basis of geothermal regulation on federal lands by the Bureau of Land Management (BLM) and the U.S. Geological Survey (USGS) is contained in the Geothermal Steam Act of 1970 authorizing leasing of federal lands for geothermal development. The regulatory authorities of these agencies differ—the BLM can decide on issuance of leases while the USGS can control activities of leased properties through sealed bid leasing.

To begin the federal process, prior to the sealed-bid leasing auction, BLM conducts an environmental study of the land. After winning a lease, the leaseholder files a Plan of Operation. As part of this plan, the USGS prepares a more detailed environmental study than the original BLM study. The USGS also issues permits for the subsequent steps of drilling and geothermal production.

For federal, state, and private lands, the final regulatory step consists of permission for commercial operation. On federal land, the USGS requires a Plan of Production for which it issues approval. In addition, on all three types of land, two final state approvals are required: The Certificate of Public Convenience and Necessity, issued by the California Public Utilities Commission, and the Approval of Power Plant Siting and Construction by the California Energy Resources Conservation and Development Commission (Energy Commission). These two approvals are issued after an electrical pricing agreement has been reached between the developer and utility company involved.

With such a large number of regulatory hurdles, it is not surprising that, recently, laws and agency regulations have been passed to speed up or eliminate steps in the process. A state law (AB 2644) enacted in 1978 assigns the Division of Oil and Gas as the lead agency for geothermal field development, with a 135-day variance approval time. Additionally, the Division can delegate its lead authority to any county which has enacted a geothermal element in its general plan. For a county such as Imperial, which has a sophisticated geothermal plan, the new law gives the county potential for a more active role in field development than was previously possible. It may do a better job than the Division of Oil and Gas, since its citizens are directly affected by adverse effects. Another streamlining of regulations recently enacted is the energy Commission's shortening of the process of Approval of Power Plant Siting and Construction. This process is now of 12-month's duration instead of 18 and one stage instead of two stage. It is interesting to note that actual experience with the new procedure is not showing any significant streamlining in practice, as the regulatory agencies subject to the reduced time limits are able to reject any application as "incomplete," return it to the applicant, and proceed to start the time clock over again. The agencies are also able to turn down praiseworthy proposals by suggesting modifications, thereby taking as long as they want to process any application by stages.

Limiting Factors on Geothermal Development

Imperial County's geothermal resource is a large and potentially exploitable one. Earlier chapter sections have reviewed the initiation of geothermal production projects, both electrical and direct use. The size and pace of eventual development will be curtailed by several limiting factors. A limiting factor is defined as a major impediment to the realization of geothermal potential. The following list presents major limiting factors for Imperial Valley:

1. Transmission lines
2. Financing
3. Political factors
4. Cooling water availability
5. Direct use transmission pipes
6. Brine waste disposal

Although all these are considered important as limiting factors, the list is ordered in importance. However, the ultimate order of importance will depend on unknown future claims and sequences of events, as the geothermal development process unfolds.

Transmission Lines

In the first chapter section, the history of the conflict over transmission lines was reviewed. Thus, the present discussion is shortened to several comments on limiting factor aspects.

Central to transmission lines as a limiting factor is the existing IID network, which can only transmit a maximum of about 300 MW_e out of the county. Although the problem of geothermal energy will be solved by construction of one or more 500-kV lines out of the county, there are delays of five to ten years in planning the corridors, purchasing and condemning land, overcoming regulatory and legal delays, and constructing the transmission lines.

Figure 8-4 presented one solution proposed by the Imperial Valley Action Plan to allow export of 2000 MW_e of geothermal energy. In this solution, the existing IID network has been expanded into a triangular-shaped collector system consisting of 230-kV lines. This collector system has nodes in all four KGRAs and has three interconnection substations to the 500-kV lines. To realize such a full system, however, will take from 5 to 20 or more years. Recently, SDG&E announced construction plans for a 500-kV line from Arizona to San Diego, to follow either the lower route or middle route shown in Figure 8-4. Construction is planned in two stages, with the Imperial Valley–San Diego link going in a year or so before the Valley–Arizona link. Energy imported to California from Arizona and New Mexico will be largely coal fired rather than nuclear, because the Palo Verde nuclear energy is already under contract. The plan is conditional on approval by the California Public Utilities Commission. This announcement appears to imply an increase above the present 300-MW limitation in the late 1980s.

Financing

Geothermal exploration, development, and production requires large sums of money. Obtaining the money so as to satisfy production companies, utilities, construction concerns, the federal government, and investors of various types becomes a difficult and complex process. As a result of the difficulties in geothermal financing, it is identified as the second limiting factor.

Geothermal Development Update

The sheer size of national geothermal financing requirements was pointed out recently by Otte (1978). For a U.S. geothermal capacity of 20,000 MW$_e$, he estimates that at least 1,200 exploratory wells and 8,000 development wells will be required. At an average well cost in 1978 dollars of $825,000, a total of $7.6 billion of financing is required. For hookup of these wells to power plants, Otte estimates $4 billion in financial requirements. At a medium capacity of 2,000 MW$_e$, by the above reckoning, Imperial County's financial needs just for full field development will be $760 million for wells and $400 million for hookup. These are staggering sums when it is recalled that the assessed value of Imperial's agricultural land is currently (1979) about $1.5 billion and the Gross County Product (1972) was $447 million. The financial costs of power plants to utilize the developed fields will be even larger.

The costs of geothermal energy have been estimated differently by various investigators. In 1982 dollars, field development costs appear to lie at about $600/KW$_e$, while power plant costs are in the $1500–$1700/KW$_e$ range for flash. Costs are uncertain for binary, owing to lack of any large size binary plant.

Modern goethermal financing requires creation of a financial package satisfying all the parties involved. The financial package is usually based on the predicted consumer's cost of the energy (referred to as the bus-bar cost). This cost is then allocated back to the utility, production company, plant construction company, and other investors. Since the consumer might be paying back costs over several decades, the order in which parties are paid off is also important.

Geothermal packages are influenced by the concept of risk. As discussed previously, geothermal field production is very risky, and geothermal reservoir engineering is a relatively new science which can only control risk to a limited extent. There are threats of temperature, pressure, and/or permeability declines, causing shutdown of part or all of a producing field.

Although field production is highly risky, the technology of power plant construction and electrical transmission is well worked out. It is logical that utility companies, which are involved with power plant construction, would be somewhat unwilling to assume the much greater risks arising from field production. There is a second reason for such hesitancy. U.S. utilities are very heavily in debt. Thus, they cannot risk a lowering in credit rating through risky geothermal losses. One utility executive (Richards, 1978) has recently estimated that an average utility company board of directors would not be willing to expose more than 10% of company earnings to losses from medium- or high-risk projects.

To place this in the context of Imperial County, 10% of SDG&E's 1975 earnings was $2.6 million, and 10% of SCE's 1975 earnings was $18.7 million. These figures show that size of utility becomes highly important in financial options. In a prior section, it was seen that financial size and strength is important also to options available to developers. Magma Power Company, for example, has such a strong balance sheet that it was able to finance its 11-MW$_e$ pilot plant entirely with internally generated funds.

Another important aspect of a geothermal financing package is taxation. Taxation or deductions may take place at the federal, state, and local levels. At the federal level, the main form of taxation is the individual and corporate income tax. The National Energy Act of 1978 increased the deductibility of geothermal investments by instituting the intangible drilling deduction and percentage depletion allowance for all forms of geothermal energy, except geopressurized, which has its own set of tax rules. The above deductions are without the restrictions which apply for oil and gas.

At the state level Wagner (1978) has categorized geothermal taxes as follows:

1. State income taxes
2. Ad valorem (property) tax
3. Direct taxes on production (severance taxes)
   a. production taxes
   b. conservation taxes
   c. privilege, occupational, or license taxes
   d. sales taxes
   e. excise taxes

In California, both corporate and individual state income taxes apply to geothermal income. In some cases, these taxes may be lessened by a depletion allowance which applies to extraction of minerals, such as salts, from geothermal operations. The property tax in California is based on the present value of a future stream of geothermal profits. It is applied about the time of actual production. Production taxes are computed at the well head or place where a product is sold.

A goal in any geothermal financing package is the reduction of risk—to some or all parties. A response to this need from the federal government is DOE's Geothermal Loan Guarantee Program (GLGP). This program involves a guarantee by the federal government for payment of 75% of an approved geothermal investment. The remaining 25% of investment funds must be obtained from private sources, such as investors. To apply to the GLGP, an applicant must submit information in the following areas:

Geothermal Development Update

1. Borrower's management
2. Geothermal resource
3. Environmental data
4. Marketability
5. Project plan
6. Financial information
7. Security
8. Legal

The first GLGP loan to the approved was for Republic Geothermal's drilling operations on the East Mesa. The GLGP has been criticized as discriminating against large companies. The reason is that any type of default, including a GLGP default, will trigger a lowering by banks of an entire company's bank ratings. For a large company, the gain from possible GLGP compensation is not worth the loss of its bank ratings on *all* portions of the company. Therefore, large companies seek entirely private financing for geothermal projects. Nevertheless, the GLGP, aiming at risk reduction, has already spurred large and significant projects. The discrimination against large companies was mandated by Congress to encourage introduction of new, independent geothermal companies not controlled by the major oil companies.

In developing geothermal financing, there are several alternative package types including the developer–utility contract, the developer–IRAC–utility contract, and the leveraged lease.

In a developer–utility contract, the developer assumes the responsibility and financing for field production, while the utility assumes responsibility and financing for power plant construction, transmission corridor construction, and electrical sales to the consumer. A contract is written dividing and allocating back to the parties the consumer's payments for electricity.

A second type of contractual agreement is a developer–IRAC–utility contract. An Interim Risk-Assuming company, or IRAC, is defined as a company which purchases the geothermal resource at the well head, converts the resource into electricity, and sells the resource to the utility at the bus-bar (Rodzianko, 1978). The IRAC falls right between producer and utility. It serves as a banker, in place of utility bondholders, in holding title to the power plant. It also serves as a project manager for power plant construction and operation. One advantage to IRAC type of packages is that the IRAC's corporate form and charter may be newly and specially created for the regulatory and financial environment of a specific geothermal site (Rogers, 1978).

A third form of financial package is the leveraged lease (Panawek,

1978). This type of financing is undergoing rapid increase in U.S. industrial acceptance. It has been applied extensively to computer equipment and aircraft. A leveraged lease is analogous to leasing a house. The "home owner" (equity investor in the power plant) is a specialized leasing company, such as Itel Corporation. The "mortgage owners (debt investors in the power plant) are large financial institutions, such as insurance companies, pension funds, banks and so forth. The lessee of the power plant is the utility company. An advantage of this arrangement is risk reduction by both the utility and the producer. It is especially suitable for financially weak utilities. On the other hand, the regulatory aspects of a geothermal leveraged leasing package, so far untested in the courts, are uncertain. Unfortunately, federal tax laws discriminate against this type of structure.

Besides packaging options, there are other measures to reduce risk in geothermal financing. These include shared projects and reservoir insurance. In a shared project, several utilities join together for percentage participation in a project. Following Richards (1978), the advantages of such sharing are the following:

1. Low risk to the consumer.
2. Reduction in each utility's capital exposure.
3. Choice, for transmission systems, of several participant's systems.
4. Easier acceptance of the developer–utility contract package. As a result, the plants are owned and operated by utilities.
5. Resource development and production is also available as an option.

A second risk reduction measure is insurance of the reservoir for both length and quality of production. As the geothermal energy sector grows, the insurance industry may develop insurance plans for geothermal development.†

Political Factors

Political factors include public opinion and leadership disputes. Leadership disputes were reviewed in the first section of the chapter. Potential areas of political conflict will surface unpredictably, depending on developmental events in all the other areas discussed in this book. Many recent energy events in the United States, including the

†PL 96-294 should help with reservoir insurance.

dispute over the location of the Sun Desert nuclear plant, have been limited by political factors.

Direct Use Transmission Pipes

Direct use applications of geothermal energy are limited by the distances from geothermal fields to the urban areas or industrial plant sites which use the hot water. In the western United States house heating, it is clear from Table 8–5 that a very small percent of the population lives within ten miles of the water-dominated geothermal fields. As is apparent in Figure 1-1 the major towns of Imperial County, El Centro, Calexico, and Brawley, accounting for 58.6% of the 1970 population, are either in the middle of a KGRA (Brawley) or bordering a KGRA (El Centro and Calexico). Therefore, distances from major towns to KGRA direct heat sources are *at the most,* several miles.

The Community Center and Holly Sugar direct use projects discussed earlier do not plan to utilize heat sources in KGRAs. Rather, drilling will be done within a half-mile distance from each site on non-KGRA land. Many of the other potential industrial direct uses shown in Table 8–10 are likewise within a half mile of KGRA or non-KGRA heat sources.

With such small hot water transmission distances relative to the rest of the western U.S., direct use would appear very favorable in the county. Such a picture, however, is made less rosy by the high cost of deep geothermal drilling in the Imperial Valley, and by the high cost and environmental problems of transmission pipes. Drilling costs are high because the three reservoirs near the populated areas are all at the minimum 3,000 feet deep. The cost of a production well at such depths ranges between $500,000 and $1,200,000. At other direct use locations such as Boise and Klamath Falls, drilling costs are much lower because the medium- and low-temperature reservoirs underneath these cities are much shallower—sometimes only several hundred feet in depth.

A second problem is the cost of transmission pipes. Goldsmith (1976a) cites costs (installed) for surface transmission pipes of $10,000/mile per inch of pipe diameter. For a typical hot water well, steam and water with one-way flow pipes will cost $200,000/mile. However, the cost is considerably increased by burying the pipe. The increase is due to costs of excavation, of special insulation to protect against ground water, and of special thermal expansion outlets. TRW (1977a) estimates that buried 14-in insulated pipe, with two-way flow, costs $1,290,000 per mile. Similar 10-in pipes would cost $850,000 per mile. In Imperial County direct use applications, two-way brine flow is necessary because of the necessity to reinject. Therefore, even for distances of, say, half mile maximum in the county, transmission pipe costs would be substantial.

Under such financial constraints, how could the huge Icelandic direct use facility have been built? In Reyhjavik, Iceland, for instance, the direct heating system serves 15,600 homes and apartments. The project economics are viable because of the consistency of Iceland's seasonal and diurnal patterns, and because of cost sharing achieved by a large number of consumers.

Environmental problems with transmission pipes stem from the extremely high temperatures (300–400°F) of the steam and brine in the pipes. There are dangers to humans and animals from pipe contact (Goldsmith, 1976a). In addition, for buried pipes, small, persistent leakages could lead to ground contamination, while a pipe burst would be a major hazard to the land.

Cooling Water Availability

If no environmental controls are present, water presents no problem for geothermal development. For example, at the Cerro Prieto geothermal plant south of Mexicali, there is minimal concern with environmental damage, such as salt water disposal on the surface and air pollution. In Imperial County, however, environmental controls will be strict. The Geothermal Element of the County General Plan has strictures against subsidence, damage to agricultural land, geothermal use of fresh irrigation canal water, and so forth. Therefore, cooling water requirements will need to be carefully analyzed by all power plant operators in the valley.

Before discussing specific cooling water needs and solutions, it is necessary to outline the county water system. The inputs and outputs to this system are presented in Table 8–13. The largest water input, the flow of the Coachella and All American Canals, consists of water diverted from the Colorado River at the Imperial Dam above Yuma, Arizona. The New and Alamo Rivers originate in Mexico. They are highly polluted on entering the United States. Both rivers collect Mexican irrigation wastes, and the New River collects municipal wastes from Mexicali.

The major outflows from the county are into crop consumptive use (evapotranspiration) and polluted discharge of the New and Alamo Rivers into the Salton Sea. The county's ground water system is also supplied by outflow—mainly from leakages in the Coachella

TABLE 8–13. IMPERIAL COUNTY'S WATER ALLOCATIONS, 1977[a]

Source	Acre feet
Input to county water system	
New River	108,000
Alamo River	1,400
Coachella and All American Canals	3,220,000
Total	3,329,400
Output from county water system	
New River discharge to Salton Sea	413,000
Alamo River discharge to Salton Sea	615,000
Coachella Canal export	387,000
Crop consumptive use	1,504,000
Ground water outflow to Salton Sea	50,000
Ground water recharge by canal seepage	261,000
Total	3,230,000
Output unaccounted for	94,400

[a] Source: VTN (1978).

Canal. Ground water usable for geothermal development is located mainly on the southern East Mesa. VTN Inc. (1978) estimates that 21.7 million acre feet of geothermally usable groundwater are in this area, with 8.8 million usable acre feet in the rest of the county. Another large body of water supplied by outflow is the Salton Sea, which had a 1977 volume of seven million acre feet. However, because of saline water inflow (about 3,000 ppm) and evaporation from the Sea surface, the Sea's salinity is 39,000 ppm–4,000 ppm more than that of seawater.

Geothermal power plants require water for cooling purposes. As discussed in Chapter 2, (see Figures 2–7, 2–9, and 2–10), different plant designs have different water flows and requirements (cooling water consumption by water source is shown in Table 8–14). A comparison of water requirements for four types of geothermal power plants is presented in Table 8–15. For the three operating plants in the table, the cooling water source is geothermal fluid condensate. Since no make-up water is used (see Chapter 2), there is a potential for land subsidence, because more water may be removed from the geothermal reservoir than is replaced. For the binary plant design at Heber, irrigation water would be used for cooling in large quantities while the geothermal brine would be reinjected.

An important control variable in cooling tower design is the num-

TABLE 8-14. COOLING WATER CONSUMPTION BY WATER SOURCE[a]

Cooling water source	Cycles of concentration	Cooling water requirements (acre ft/MW yr)					
		Heber, Brawley, and E. Mesa			Salton Sea		
		Evaporation	Blowdown	Total	Evaporation	Blowdown	Total
Steam condensate	10[b]	75	8	83	50	5	55
Agricultural waste water	5[c]	75	19	94	50	12	62
Irrigation water	4[d]	75	25	100	50	17	67
Ground water (East Mesa)	10[c]	75	8	83	50	5	55
Salton Sea water	2[c]	75	75	150	50	50	100

[a] Source: Layton (1978).
[b] Blowdown may be discharged to surface waters or disposed of by subsurface injection, evaporation ponds, etc., depending on site location and blowdown quality.
[c] Assumes no return of blowdown to surface waters.
[d] Assumes disposal to surface waters, provided that water quality objectives and efficient standards are met.

TABLE 8–15. WATER REQUIREMENTS FOR TYPICAL GEOTHERMAL POWER PLANTS[a]

Plant type	Location	Net power output (MW)	Reservoir characteristics		Water flow through cooling towers			Water source	Water disposal
			Flow (10^3 lb/hr)	Temperature (°F)	Evaporation (10^3 lb/hr)	Blowdown (10^3 lb/hr)	Total (10^3 lb/hr)		
Hot water single flash	Ceno Prieto Mexico	75	6150	385	220	1130	1350	Geothermal fluid	Evaporation pond
Hot water double flash	Hatchobaru Japan	50	2420	574	840	104	944	Geothermal fluid	Reservoir reinjection
Hot water binary	Heber California	65	7600	360	1270	330	1600	Irrigation water	Irrigation system
Dry steam total flow	The Geysers California	53	970	460	800	170	970	Geothermal fluid	Reservoir reinjection

[a] All plants except Heber are operating. Heber figures are based on plant design. Source: VTN (1978).

ber of cycles of concentration. The salinity and water balances may be artificially controlled in a cooling tower by varying the number of times water is circulated and recirculated through the cooling elements. As shown in Table 8–14, highly saline Salton Sea water is only cycled twice, because of water quality requirements on the cooling tower waste water (blowdown). However, the price paid for such salinity control is the very high volume of blowdown. By contrast, very pure steam condensate or groundwater may be cycled 10 times, resulting in a very slight volume of blowdown. Another way of putting this is that cooling towers can handle any of the Imperial Valley waters, but the price for poor waste quality will be paid in high salinity or high volume of outflow water.

To compute cooling water availability, options for power plant/cooling tower design are charted against surface water availability in Imperial County's water system. Such a study (VTN, 1978) is shown in Table 8–16 for each of the major KGRA areas. Water availability in acre feet has been converted into the power plant megawattage supportable by the available water. The study assumed maximum field capacities of 2000 MW_e for the Salton Sea, 1000 MW_e for Brawley and Heber, and 500 MW_e for the East Mesa. The power plant option, indirect contact condensers, refers to the presence of a special shell

TABLE 8–16. WATER AND RESERVOIR CONSTRAINTS ON ULTIMATE GEOTHERMAL MEGAWATTAGE FOR IMPERIAL COUNTY KGRAs[a]

KGRAs	Cases						
	Flash				Binary		
	A	B	C	D	E	F	G
Heber	1000[b]	1000[b]	75	1000[b]	75	1000[b]	0
East Mesa	500	40	75	500	75	500	0
Salton Sea	2000[b]	2000[b]	40	2000[b]	40	2000[b]	1650
Brawley	1000[b]	1000[b]	40	1000[b]	40	1000[b]	0

[a] Cases: A, flash steam, zero make-up water; B, flash steam, complete reinjection, use of agricultural waste water; C, flash steam, complete reinjection, use of canal water, indirect contact condenser; D, flash steam, complete reinjection, use of ground water, indirect contact condenser; E, binary, complete reinjection, use of canal water; F, binary, complete reinjection, use of ground water; G, binary, complete reinjection, use of agricultural waste water. (Source: VTN, 1978.)
[b] Constraint on geothermal reservoir; water constraint not reached.

and tube heat exchanger in place of the usual direct contact condenser. The purpose of this design is control of air pollution, because indirect contact condensers prevent the release of any pollutants from the cooling towers. The results are highly favorable for cooling water availability. Power plant options A, D, and F allow maximum field development in all four KGRAs and option B (flash plant, used agricultural drain water) allows maximum development in three out of four KGRAs. Option B is constrained in the East Mesa because of the slight flow of the Alamo River in the southern part of the county (the Alamo increases its volume of flow 440 times from the Mexican border to the Salton Sea outlet). Use of canal water (options C and E) leads to very restricted megawattage because the county allows, in its General Plan, for only enough canal water for one power plant in each KGRA. Finally, option G (binary, use of agricultural waste water) is only possible in the Salton Sea, through use of the great volume of Salton Sea water.

Although the volume of cooling water does not appear to be a major limiting factor for geothermal development, this factor is complicated by the potential effects of water options on the Salton Sea. When the Sea was formed, it rose to a height of 80 feet and then receded to a height of 55 feet.† Since 1925, the Sea has been rising. If geothermal water options cause a rise, landowners on the Sea edge will be threatened by property loss. On the other hand, if an option causes the Sea to fall, sports facilities at the Sea edge will be endangered. If the Sea's salinity rises, certain aquatic species in the Sea, such as the sport fishes orange mouth corvina (*Cynoscion xanthulus*), sargo (*Anisotremus davidsoni*), and gulf croaker (*Bairdiella icistia*) will be threatened (Layton, 1978). The threat is due to egg and larvae mortality at salinities above 40,000 ppm (May, 1976; Lasker et al., 1972). Goldsmith's (1976b) computer simulations to determine effects of water options on the Sea's level and salinity tended to confirm these observations (see Table 8–17).

In summary, cooling water availability is a minor problem. However, environmental and legal ramifications of large-scale use, such as land subsidence, water pollution, and effects on the Salton Sea, may place limitation on water use. The extent of limitation is unknown, because the environmental problems cannot yet be studied by experiments and, consequently, have been estimated mainly by computer simulation.

†It should be noted that the Salton Sea is below sea level.

TABLE 8–17. COMPUTER SIMULATIONS OF GEOTHERMAL WATER OPTIONS ON THE SALTON SEA[a]

Water option	Size of plants	Results on Salton Sea	
		Water level	Salinity
Direct discharge of brine	One well	Negligible	Negligible
	100 MW (Salton Sea)	0.4 foot rise	Doubles rate of increase
	100 MW (other location)	0.4 foot rise	20% higher rate of increase
	1000 MW	Over 10 foot rise	
Use of irrigation drainage	100 MW	0.3 foot rise	+350 mg/l[b]
	1000 MW	2.5 foot fall	+350 mg/l[b]
Reservoir injection of sea water	2000 MW	4 foot fall	Safely stabilized in 40 years

[a] Source: Goldsmith (1976).
[b] Current sea salinity is about 35,000 mg/l.

Brine Waste Disposal

A final limiting factor is the disposal of geothermal brines. Geothermal brines are very hot and corrosive. In addition, they may contain potentially damaging biological materials. There has been little experience worldwide with *environmentally regulated* brine disposal. The reason is that many existing plants are permitted to cause significant pollution from brine wastes. At Cerro Prieto, for example, brine wastes such as those in Figure 8-7, are discharged into an evaporation pond. At Lardarello, Italy, most wastes are disposed, untreated, into local streams, while at Wairakei, New Zealand, all brine wastes are emptied into the Waihato River.

What brine disposal options are available? Differding and Walter (1978) propose the following list:

1. direct surface discharge
2. treatment and surface discharge
3. ponding
4. secondary use of effluents
5. reinjection
6. reinjection with pretreatment

Options 1 and 3 appear ruled out in Imperial County by the County General Plan. Presently, option 5 is the most popular one in the planning of power plants in the county. However, this option has several major problems. First, there may be plugging of the reinjection wells. Second, there may be reduction in the permeability of the geothermal field due to precipitation of the reinjected brines.

Options 2, 4, and 6 have the disadvantage of requiring sewage treatment facilities. The cost of these options will be related to the chemical quality of the geothermal brines. The extent of limitations from the brine waste disposal factor is unknown for a water-dominated resource in an environmentally regulated region. Presently, no provision has been made by the county for a shared disposal site. Although most individual developers appear to favor option 5, they may be forced into forms of water treatment. Republic Geothermal has indicated that they feel treatment is essential for their projects in the Imperial Valley.

Conclusions

This chapter has discussed the present status of Imperial County's geothermal development. The key indicator of extent of development in any geothermal area, unless direct use predominates, is electrical generation capacity.

Much exploratory and regulatory groundwork has been completed by energy developers for the initial 300–800 MW_e of geothermal capacity which will be installed in the 1980s. Capacities beyond this level are dependent on the extent and timing of installation of high-voltage transmission corridors. The opening in 1979 of Magma Power's 10-MW_e plant signals the initiation of generating capacity, which may total in thousands of megawatts by early in the next century.

In earlier chapters, the potential for geothermal conflicts was discussed. Already, several years before the advent of large-scale production, conflicts are emerging—disputes over transmission corridors, the training/unionizing of geothermal labor, and location of the Brawley sewage plant. These disputes are related to county characteristics discussed earlier—levels of leadership, the county's occupational and employment distributions, and juxtapositions of population and geothermal energy. As development unfolds, many other disputes will surely emerge. As will be discussed in the final chapter, county leadership and citizenry will benefit by the broadest possible model for development.

Direct uses of geothermal energy, also called nonelectrical uses, are being generated in the county. Major projects underway are the El Centro Community Center and Holly Sugar plant. The Valley Nitrogen plant has the potential for application of direct use. Many other industrial, commercial, and residential uses are being discussed, given the limitations of the resource depth and cost of surface transport.

Limiting factors will restrict the extent and timing of resource development. Transmission line locations and capacities will constrain electrical capacity. The potential exists for expanded local and extralocal conflicts over corridor locations, environmental effects, and funding. The financing factor is the result of the large sums of money required, and how to obtain such large sums relative to the accompanying financial risk.

Political factors may become important as limitations, if public opinion conflicts and leadership splits occur.

As limiting factors, direct use transmission pipes, cooling water supply, and brine waste disposal will vary in magnitude depending on environmental restrictions and concerns and on the economic costs of various options.

In all, geothermal development in Imperial County is progressing, but not without problems. There are needs for further research and education, that as many facets as possible of the development process may be anticipated before they occur.

9

Research Conclusions and Policy Recommendations about Geothermal Development

INTRODUCTION

Much of this research has implications and generalizability for other areas potentially undergoing geothermal or other energy resource development. The immediate goal of the overall study,† however, was to provide Imperial County with information which would be helpful in the drafting of a geothermal element for their *General Plan* (1977). Thus, the county desired a relatively broad investigation giving them knowledge about the resource base and potential problems associated with its development. The environmental segment of the study was designed to acquire an integrated understanding of the natural and artificial landscapes giving Imperial County its character and that will play a role in resource development. Thus, integration, interdependence, and synergism were key words, while detailed individual analysis was given a lesser priority.‡

The population study was designed to analyze the demographic character of the county especially in terms of proximity to Mexico, the rural nature of the surroundings, education level, and labor force composition. The public opinion and leadership surveys were undertaken to evaluate local interest and knowledge of geothermal energy, gauge

---

†Besides the specific studies discussed so far, the contract also supported research on land use, politics, legal and institutional aspects, engineering, economics, and geophysics.

‡Lawrence Livermore Laboratory also has conducted a detailed environmental analysis of Imperial County (Layton and Ermack, 1976; Phelps and Anspaugh, 1976).

feelings about its development, especially *vis à vis* agriculture, and to help establish appropriate regulations.

Major Research Conclusions

Environmental Impact

For all practical purposes, environmental impact from geothermal development in Imperial County will be concentrated in the most agriculturally intensive part of the county—the Imperial Valley trough.† This artificial landscape of fields, feed lots, irrigation and drainage networks, the Salton Sea, highways, and residential communities overlies the majority of the county's known geothermal resources. The agricultural valley is central to virtually all environmental issues and thus plays a key role in geothermal development; concern in the valley about agriculture has already caused some delays. Thus, while concern for the primitive, pristine, or isolated nature of an environment has often slowed energy development, ironically, the completely remade Imperial Valley may pose a substantial obstacle to geothermal development.

One of the greatest concerns in this area of delicately leveled irrigation and drainage channels is possible subsidence from geothermal fluid withdrawal. Official county policy requires that agriculture not be harmed by geothermal activity. Thus, substantive damage from subsidence could hinder or even permanently forestall geothermal development. Subsidence computer models have been derived for the valley (Finnemore, 1976) and first- and second-order survey networks have been established for baseline and monitoring purposes (Lofgren, 1974). Neither has yet led to definitive conclusions of risk potential. Monitoring is complicated by natural subsidence and upward movement. Thus, baseline data must be collected for several years prior to development before any subsidence could be attributed to geothermal development. To help monitor local elevation changes, all geothermal facilities in Imperial County are required to be tied into the existing geodesic network. It is apparent, however, that conclusive answers must await commercial sized plants [50 MW_e+] with their continuous, substantial, and long-term fluid withdrawal and injection.

It is an unusual coincidence that the geothermal resources of Im-

†Present geothermal development on East Mesa is technically outside of the irrigated area, but it is not clear whether development there will or will not have an impact on agriculture.

perial Valley underlie an area so sensitive to disturbance by subsidence. Wairakei, New Zealand is the only other geothermal site to have experienced substantive subsidence, but even several meters of displacement there have hardly affected operations (Stilwell et al., 1975). For other areas subsidence should not be as much of a concern, but even when it may be, experience from the Imperial Valley will add only nominally to the scientific experience, and those new areas will also probably have to await larger-scale commercial operations for a proper and useful evaluation.

Based on experience elsewhere (Evans, 1966), another matter of environmental interest is that of possible seismic activity induced by the withdrawal of hot fluids or the reinjection of waste fluids (Hill et al., 1975). Preliminary theoretical calculations made for heat withdrawal in Imperial Valley indicate that a maximum potential earthquake would be no greater in magnitude than has been experienced from natural causes in the past (Lee, 1977). Moreover, fluid injection seems to induce seismicity only rarely, and has even been suggested as a means of seismic control (Raleigh, 1976). Even so, any increase in noticeable seismic activity in Imperial Valley would be received unenthusiastically, and no one can accurately assure its absence. As with possible subsidence, then, apprehension regarding induced seismicity cannot be alleviated until commercial-sized units are in operation.

Air pollution, particularly noncondensible gas emissions such as hydrogen sulfide, has impeded geothermal development at The Geysers field in Northern California (Castrantas et al., 1976; Reed and Campbell, 1976), and concern for its impact on Imperial Valley crops has elicited detailed study (Shinn et al., 1976). The prospect for a repeat of The Geysers experience in Imperial Valley is unclear. Good results are predicted for the operation of new scrubbing equipment at The Geysers (Castrantas et al., 1976; Semrau, 1976; Allen and McCluer, 1976), and experiments with hydrogen sulfide have indicated small, and in some cases, beneficial effects on some crops grown in the Imperial Valley (Shinn et al., 1976; Thompson, 1976). On the other hand, Cerro Prieto, an operating geothermal facility just south of Imperial County in Mexico, see Figure 9–1, has unabated H_2S emissions 2.6 times greater by weight than for The Geysers' Units 1–10, which do not have abatement devices (Axtmann, 1976).

Much of the concern registered at The Geysers' operations is a result of the immediate downwind proximity of homes and summer resorts. The odor concentration of sulfur (30 ppb) has had a notable impact in arousing nearby residents, who have in turn demanded action from their elected officials. In addition to an interest in health and

FIGURE 9-1. Steam ejection, production well, Cerro Prieto, Mexico.

aesthetics, air pollution in Imperial County may also effect crops. Part of the interest in all these effects has been a spill-over from The Geysers. Public education programs have been suggested for both locations, and some educational efforts at The Geysers has had some impact in assuaging residents' fears. Nevertheless, resort owners believe their guests might not be so understanding. A lowering of H_2S concentration below the odor threshold (and, incidentally, the state standard) by the installation of appropriate scrubbing devices seems to be the solution to this problem in both locations.

One other possible effect of geothermal development on air quality is climatic change. Kelly (1976), in a study based on radiosonde data from the Yuma Proving Grounds, 80 km east of the valley, predicted no appreciable effect on climate from the operation of a 50-MW_e power plant.

Geothermal plants, pipelines, and channels for hot source fluids and wastes, access roads, pads for production and injection wells, and transmission line rights of way will require some agricultural land be taken out of production. Calculations discussed in Chapter 5 indicate that the amount of land actually displaced by a 10,000-MW$_e$ geothermal program will be 1% – 2% of the total cultivated area of the county, and probably closer to the lower figure. One percent of Imperial County land is about 5000 acres, a loss which could be partially offset by the cultivation of other, presently idle, somewhat less fertile land (assuming available water). Five thousand acres may be contrasted to somewhat over 42,000 acres for the same 10,000-MW$_e$ coming out of a coal-fired power plant.†

At the hilly Geysers' site, the comparison would not favor geothermal energy so strongly, but such terrain is not the rule at other geothermal sites (e.g., Mono Lake–Long Valley and Coso Hot Springs in California; Brady Hot Springs in Nevada; and Roosevelt Hot Springs in Utah). In comparison with typical fossil fuel plants, land use considerations should not be an obstacle to geothermal development; however, the concentration of activities, even though requiring less total land, can present some problems when the land under development is valuable (as in the Imperial Valley) or hilly (as at The Geysers). In either area, monetary returns from energy development should outstrip those from existing land use.

Because of relatively low fluid temperatures, geothermal operations everywhere suffer, particularly low efficiencies. In Imperial County, such efficiencies are not expected to exceed 12% (compared to 40% for a new coal-fired power plant). However, these figures do not compare overall production which includes extraction and transport costs. Low efficiency in power production translates into high cooling requirements, portending a problem for geothermal developers in arid lands such as Imperial County. The Colorado River allotment to Imperial County is over 3 million acre ft annually, but it is fully committed, primarily to agriculture. This water is dispersed throughout Imperial Valley by a series of irrigation canals, Figure 9–2.

†This chapter is modeled after the Navajo power plant at Page, Arizona and assumes: 494 ha for powerplant, 310 ha for ash disposal area, 2914 ha for 30 years of mining at current extraction rates at the Black Mesa Coal Mine, and 184 ha for the railroad right of way between the mine and the powerplant. This is exclusive of powerlines. If powerlines were included (assuming 800 km between the Navajo plant and the San Diego-Los Angeles load center and 240 km between Imperial County and the same load center), the comparative numbers would be about 53,000 ha for coal and about 13,000 ha for geothermal.

FIGURE 9-2. Irrigation canal, Imperial County.

The allotment is unlikely to increase in the future. Except for some short and small initial operations, cooling water will not be obtainable from high-quality irrigation supplies but rather will come from the otherwise unusable agricultural return water found in the New and Alamo Rivers, or from other sources (See Table 8–17). Using wet cooling towers, sufficient volume in these rivers exists for cooling a total capacity in excess of 10,000 MW_e (Goldsmith, 1977). In order to provide constant availability of cooling waters, river flow will have to be impounded if river flow becomes highly erratic or dries up owing to a combination of annual weather, runoff, and agricultural factors combined with geothermal withdrawals. As seen in the last chapter, cooling water demand varies greatly by power plant design, by desired salinity level of cooling tower effluent, and by location; however, cooling ponds are a necessity in geothermal power production, Figure 9–3. Such considerations would apply to the southern portion of the Alamo River, which has a very low volume of flow. The East Mesa KGRA would be affected. Otherwise, only diversion structures will be needed.

The question of cooling water supply also has several implications for the Salton Sea. The Sea receives 80% of its inflow from Imperial County wastewater. Usually, this inflow about matches evaporation. If flow of the New and Alamo Rivers was diverted for geothermal power plant cooling requirements, the level of the Sea could drop. Because initial geothermal operations will be small scale, the level of the Salton Sea will not be affected by early geothermal energy development. However, longer-range effects are unavoidable if the supply,

Research Conclusions and Policy Recommendations 311

methods of delivery, and use patterns of Imperial County water remain the same during large-scale geothermal development.

Curtailed discharge would threaten the recreational value of the Salton Sea. Dissolved solids in the Sea, now about 38,000 ppm, have been increasing at an average rate of about 550 ppm per year. "Above 40,000 ppm it is likely that reproduction of corvina and other sport fishes will decline. Then, as the salinity continues to increase, the fishery will gradually or suddenly die out as mature fish eventually succumb to the rising salinity. Other forms of water-oriented recreation and wildlife uses provided by the Sea will also become greatly diminished as the salinity increases" (U.S., 1974).

Since the Sea is the locale of over half a million recreation days per year for hunting, fishing, boating, water skiing, swimming, camping, picnicking, and nature study, increasing salinity and/or the actual evaporation of the Sea itself will not be met with favor by recreationists or others who enjoy or rely on this vast desert lake. There are other possible nongeothermal events such as imposition of double impoundment, which may also affect the Sea in the future. Whatever the chain of events, a long-term solution to the Sea's problems will be forestalled until the actual effects from geothermal development have

FIGURE 9-3. Cooling water evaporating pond, Magma Power's pilot plant, East Mesa.

been established by long-term commercial operation. More efficient use of the water supply may represent a compromise solution acceptable to everyone, but substantive work on this possibility, other than lining some irrigation channels, remains untried.

As perplexing and unresolved as the cooling water problem is in Imperial County, so far it has not been a serious threat to the development of geothermal energy in other areas. While much of the greatest concentration of geothermal resources has been identified in the arid and semiarid sectors of the country, few of those areas currently are under cultivation. Nor have they the peculiar sensitivity to subsidence plaguing the agricultural sections of Imperial County. In such areas where condensate can be used for cooling, less than total reinjection can proceed without risk that subsidence could ruin a profitable enterprise.

In addition to the threat to water quality from decreased inflow, the Salton Sea could be contaminated by overland flow of highly saline geothermal fluids into the Sea. For reasons of subsidence prevention and the maintenance of reservoir temperature and pressure, however, waste fluids will be reinjected, initially reducing potential pollution of water discharging into the Salton Sea.† Blowouts and other surface leaks would be a serious threat to crops. Prevention devices and careful monitoring have kept the global occurrence of well blowouts infrequent and usually shortlived (National Petroleum Council, 1971). Only a handful of long-term incidents have occurred at operating geothermal fields but the possibility of such an occurrence in agriculturally important Imperial County is a sensitive issue whose prevention cannot absolutely be assured. Surface leaks from shearing of wells by seismic movements and other causes are added concerns, although secure technical solutions are being sought.

For reasons of productivity, agricultural land use, urban planning, pilot safety, and other factors discussed in Chapter 8, the location of future electrical transmission lines has generated concern. For a geothermal collection network, land removed from production by transmission lines, however, will be small (less than 6600 ft^2 per mile). Moreover, pilots engaged in aerial spraying have indicated that as long as lines for a collection network are positioned in excess of 40 ft in height and located on property lines, they will be able to avoid them and still make their applications properly. High-voltage lines to

†Two published accounts relate favorable injection tests (Otte, 1970; Chasteen, 1974), but recent fluctuations in injection pressures at the DOE-SDG&E test facility would have to be stabilized before injection there could be conducted commercially and in an environmentally safe manner.

Research Conclusions and Policy Recommendations

transport energy out of the county, however, will consume much larger land acres and will cause more severe aerial problems from the 300-ft-high transmission towers.

In sum, it is the specific agricultural nature of the Imperial County landscape (and the economy it supports) which has been central to raising and maintaining local environmental consciousness about geothermal development. This awareness notwithstanding, investigations have so far uncovered no environmental impact warranting curtailment of initial geothermal development. Development, nonetheless, still faces several environmental unknowns. The most potentially significant problems are closely tied: subsidence, seismicity, water supply effects, and biological impact. If the threat of subsidence does not develop, concern for the others will be reduced substantially. Evaluation of subsidence and related problems must await collection of several years of operational data to allow comparative analysis with baseline data now being gathered through several research programs.

Population and Economics

The fundamental long-term population-economic characteristics of Imperial County have been an agricultural orientation and an ethnic composition of 46% (in 1970) Spanish-American (SA). The population analysis reported in Chapters 3, 4, and 5 identified several county features likely to be of importance in the geothermal development process.

Migration trends are important because they have been of great significance in determining the current age structure and total county population. From 1930 to 1970, Imperial County experienced a net out-migration of about 0.7% per year. Out-migrants have tended to be young adults, resulting in loss to the county of many potential labor force entrants, some with high skill levels. This steady out-migration is a reflection of a weak overall county economy relative to nearby counties, as well as a lack of economic diversity limiting career opportunities. While substantial numbers of persons have been recorded as entering the county (14.4% of the total population for the period 1965–1970), large numbers also are leaving. Thus, the turnover of individuals is much higher than net migration indicates, with perhaps 40% of the population involved in intercountry moves from 1960 to 1970.

Three interrelated variables, youthful age structure, percent SA, and fertility, were crucial in the population research. These variables were determined to be interrelated as follows: A high proportion SA is correlated with high fertility; SAs have fertility rates about 40%

higher than anglos. A well-established demographic law is that, under steady migration rates over an extended period, high fertility creates a significantly younger age structure. This triad of variables is important for population projections of county employment. The combination of high fertility and a young age structure generally leads to rapid population growth, as evidenced presently in many underdeveloped countries. In Imperial County, potential for rapid increase has been held in check by persistent out-migration.

In our study, the unemployed were analyzed by examining three subcategories, the product of which is the ratio of unemployed to total population: (a) age eligibility (number of persons aged 16+/total population), (b) labor force participation (number in the labor force/number of persons aged 16+), and (c) unemployment rate (number unemployed/number in the labor force). Geothermal energy development may lead to lower fertility due to a lower percent SA from in-migration of non-SA workers. Lower fertility, in turn, should lead to an older age structure. Thus, under the assumption of significant non-SA in-migration, there is a likely tendency that geothermal development will produce an older age structure, would increase age eligibility, and possibly could increase the proportion of unemployed to total population. If, after geothermal development, the high out-migration continues, the present proportion unemployed would tend to remain unaltered.

Another important factor related to geothermal development is the labor force skill mix. Based on 1970 U.S. Census employment and industrial classification data, Imperial County had lower skill levels than the California labor force. The most significant cause of this difference is the large component of agricutltural farm laborers (25.1% for males in the county as compared to 4.2% for males in California). This contrast may be even larger since there are from 6,000 to 12,000 commuters, not included in these statistics, who cross the border daily from Mexico and work almost entirely as agricultural farm laborers (U.S. Senate, 1971). However, even ignoring farm categories, Imperial County has only 22.2% of its labor force in professional/managerial categories, compared to 26.7% for California.

Geothermal construction and operation, even at full development, requires relatively few workers. However, skill level requirements will be high and the necessary work force probably will come from outside the county (Rose, 1977). More jobs will be created, particularly if local industrialization is stimulated as an offshoot to geothermal electrical and/or nonelectrical development.

A valuable related survey study on potential, generalized, industrial plant siting showed that labor force quality ranked first out of 43 variables in decisions by industry on plant siting (Shively, 1974). On the other hand, energy variables were ranked fifth, ninth, and twelfth as a basis for decisions. Hence, the current low skill levels of the labor force, which stem in part from out-migration of more skilled youth and lack of economic (and employment) diversity, may result either in curtailed industrial development or reliance on skilled outsiders induced to migrate into the country.

Another unknown in assessment of future skill levels is the large and rapidly growing population centered in the Mexicali urban area, which was over 600,000 as of 1975. Border regulations between the U.S. and Mexico have been notably changeable and unpredictable (Samora, 1971). For example, in the 1960s, large numbers of Mexican workers, mostly unskilled farm laborers, were allowed to work and reside in the county. During the geothermal development process, which will begin in the early 1980s and continue for decades, such regulations might be liberalized from their present strictness to make use of the potential Mexican labor pool in Imperial County industries—a pool which might become as large as a half million and include substantial numbers of highly skilled workers.

Since geographical proximities are so vital for geothermal energy development, the regional study of socioeconomic characteristics discussed in Chapter 3 was performed. This part of the research revealed many unusual regional features within or bordering KGRAs, such as concentrations of Spanish language residents in Calexico and concentrations of males with high age eligibility in Brawley and in certain districts in the central and southern portions of the county.

Another geographically oriented segment of the research involved a study of the potential farm labor force reduction resulting from geothermal development. The numbers of farm laborers likely, with present agricultural technology, to be displaced were calculated based on assumptions about future county geothermal megawattage, power plant land consumption per megawatt, and number of farm laborers per hectare of land consumed. Projections of megawattage were taken from the middle projection series of a prior study which reached a maximal county level of 3500 MW$_e$ in the year 2010 (Davis, 1976). Land consumption was projected by estimating interstitial land reduction (I.L.R.), which was defined as the percentage reduction of land in the entire geothermal well-siting area, not including land immediately surrounding central power plant and well sites. The maximal I.L.R.

assumed for the county was 35%. With maximal assumptions, only 2% of the 1970 county farm labor force is estimated to be displaced by geothermal development (Johnson, 1977).

Population projections were used to estimate the potential for growth under a variety of assumptions. In accordance with standard projection methods, these were not intended to determine definitively what will happen in the county, but rather to answer the question, "If certain assumptions are made, what populations might be expected in the future with geothermal development?" The projection series reported in an earlier chapter, using 1970 and 1975 Census and state vital statistics, records as input into modified standard population projection routines. In all series, fertility and mortality were assumed constant at the rates for the initial year (either 1970 or 1975). Migration was varied by assuming no net migration (series I), historical county rates (II), rates linearly scaled up from historical levels in accordance with linear additions to plant capacity to an eventual rate equal to 20% of the average rate for fast-growing California counties 1950–1970 (III), and a series identical to III, except for a final maximal rate equal to 30% of fast-growing California counties 1950–1970 (IV), rates scaled up as in series III and IV except that the additions of geothermal plant capacity are assumed to follow Davis' (1976) medium scenario in capacity additions and migration increase is lagged 10 years behind capacity additions (V and VI). The latter four projections are not intended to reflect geothermal energy development alone, but rather roughly to reflect a complex of geothermal, geothermal-induced, and nongeothermal concurrent trends. Examples of possible concurrent trends are a potentially large national fertility increase and the continuation of a national trend of return to rural areas.

Year 2020 projection totals for series I–VI are 175,081, 71,195, 231,851, 136,887, and 150,105 and 292,570, respectively. Since the assumptions of series I, III, and VI are not extreme, the large potential for population increases in the county is revealed.

County populations over 175,000 imply significant changes from the steady population total over the past 40 years of from 60,000 to 70,000. An important question of distribution arises from such potential population increases. Where will the additional population be located in relation to geothermal fields, since nearly all urban populations in the county currently border KGRAs? In planning for the location of possible new residents relative to geothermal pollutants and hazards, use of the environmental conclusions of this and other studies may mitigate adverse development.

A stepwise discriminant analysis established the variables distinguishing county KGRA regions in 1970 from each other and from nongeothermal areas ("stepwise" refers to ordered choice of independent variables based on the extent to which the variance in the dependent variable is explained). Of the variety of independent variables chosen for analysis, the key ones were utility gas and percent SA, both closely related to urban–rural gradients as defined by the census. Since KGRA areas differed socioeconomically from each other and from non-KGRA areas, effects of geothermal operations (e.g., noise, smell, nonelectrical development, etc.), which also differ between these areas will have selected, socioeconomic impacts. These selective impacts may be important to planners, regulatory agencies, and others concerned with more localized types of geothermal effects. However, whether such differences will persist in the decades after geothermal power comes on-line is unanswerable at this time.

Potential employment effects from prospective geothermal development were studied by stepwise regression analysis using the 1970 enumeration district data. Dependent variables were age eligibility, labor force participation, and unemployment. Independent variables included ethnicity, fertility, housing density, and economic factors of income, housing values, and rental levels. The regression analysis revealed the importance of higher fertility (negative effect) and lower density (positive) for age eligibility, of higher income (positive) for participation, and of fertility and economic variables for unemployment. If the county remains relatively stable (population projection II), these results have several important implications for energy development. In general, geothermal influence on the county fertility pattern by an increasing urban proportion or by high proportions of anglo inmigrants may result in an older age structure. This, combined with lower fertility and increases in personal income resulting from geothermal development, may affect population increases from the present low levels of employment relative to total population.

In the Geothermal Element, the county added the following two implied research results which were not actually a part of our research inputs:

1. Nonelectric utilization of the geothermal resource will have the greatest long-range employment benefits.
2. Population "Projection III appears to agree with other projections and probably is the one most likely to occur" (County of Imperial, 1977; 39).

Since most of our research conclusions on population were incorporated into the County General Plan, we will discuss the items which were excluded from the plan and comment on the relevance and importance of these exclusions. Five research results from the population analysis were excluded in the following four areas:

1. *Unemployment.* The exclusion of a conclusion about possible decreased unemployment would seem to be an oversight by the county as the county ostensibly wishes to encourage public acceptance of geothermal development.

2. *Upgrading in skill levels from geothermal in-migration.* While upgrading of skills from geothermal related in-migration appears to be beneficial to the geothermal development process and hence to the county economy, the exclusion of these conclusions may be due to employment threats felt by current residents from the potential arrival of additional skilled workers. However, inclusion of these conclusions in the plan might assist in activating programs for less skilled local residents to upgrade their skills before development.

3. *County population distribution.* The exclusion of our conclusion about local effects may be due to the county's lack of familiarity with the data base and computer mapping technologies used in this research field. However, since geothermal energy is inherently regional in character, it would be wise for the county planners to familiarize themselves with the concepts of population distribution at an early stage in geothermal development.

4. *Population projections.* The county added its own interpretation to our population projections. These projections were presented by the researchers as six alternative pathways for county population, dependent on alternative assumptions. However, we did not indicate a preference for any particular pathway.

The county indicated projection III would be the series most likely to occur (see Chapter 5.). If they wished to make such a preference, it would have been preferable to note their divergence from our research.

Policy Recommendations of the Population Analysis

We made the following policy recommendations to the county stemming from the population study:

1. Initiation of a retraining program to upgrade the skills of current residents in geothermal-related job areas.
2. Encouragement of skilled out-migrants to return.
3. Encouragement of creative regional geothermal planning. This recommendation is based on the wealth of detailed data avail-

able on county geothermal sites from the geographical and population portions of the NSF/DOE project, as well as from a large concurrent study conducted by Lawrence Livermore Laboratory on the environmental aspects of geothermal energy in Imperial County.

Of these recommendations, the following two were excluded from the General Plan:

Encouragement of Out-Migrants to Return. Such a policy recommendation was likely considered unconventional for a local government. Also, the county would likely be handicapped by poor record keeping, although such encouragement is common practice in the private sector.

Encouragement of Creative Regional Planning. The following sentence from the General Plan gives some mention to regional planning: The county should "undertake studies to determine the location of and the services required by the increased population and any other effects engendered by geothermal development." This recommendation, however, did not go far enough in suggesting a modern regional information systems approach. Such an approach would likely pay dividends in surmounting the entanglements of regulations which will be present during the energy development process.

Public Opinion

A public opinion survey of Imperial County residents in 1977 using both structured and open-ended questions showed almost 90% in favor of development. However, three out of every four people believe that geothermal development should be strictly regulated. Less than 10% felt that geothermal development could be completely unregulated. About three out of every four people in Imperial County felt that preserving the environment was an important issue; however, only about one out of every five people saw any environmental or social problems arising out of geothermal development. For those people who thought there would be problems, most of them felt that there would be environmental degradation as opposed to problems such as population increase, need for more services, tax increases, and so forth.

The most important response clusters to an open-ended question about expectations from geothermal development were the following: (1) about one out of five persons believed that it would result in cheaper energy; (2) a little under 20% felt that it would bring more power to Imperial County; and (3) about 15% mentioned a variety of miscellaneous positive benefits. Other less often reported expectations were more industry and jobs and minor environmental changes. Only approximately 5% of the population had negative expectations.

The small population segment who did not care or were opposed to geothermal development were more likely to have only an average or less than average understanding. Thus, level of understanding of geothermal resources is related to favoring or opposing it. Similarly, knowledge about geothermal resources increases one's belief in the need for strict regulation.

Rural landowners in Imperial County generally accept the idea that growth is good for the community. Indeed, this acceptance of the need and desirability for growth permeates survey responses. Leaders and citizens are in accord. While environmental protection engendered little interest on the part of many community leaders, most county officials and irrigation district spokesmen, as well as the general public, clearly expressed an extensive environmental concern and a desire for strict regulation. On the other hand, there appears to be little interest on the part of the community leaders and county officials in developing citizen participation in major decisions, e.g., geothermal development.

A lack of interest in developing citizen input probably is reflected by the small percentage of respondents who believed that citizens of Imperial County have been adequately informed about geothermal development. Thus, wider participation by citizens is viewed by some community leaders as an encouragement for conflict and controversy. Yet, responses by community leaders as well as citizen responses indicate that there are few overt issues currently dividing Imperial County residents.

While voting patterns outwardly seem to reflect a consistency in responses of leaders, citizens, and (particularly) the political arena, there are undertones of potential conflict. Issues have arisen over the past few years threatening the consensus. This potential for conflict could erupt in the future over geothermal development. The leadership may not be cognizant of the fact that, while citizens are almost universally in favor of geothermal development, they also are almost as extensively in favor of strict regulation. Similarly, only rarely does a person in Imperial County express the opinion that citizens have been adequately informed about geothermal development.

It can be safely assumed then that, if geothermal development proceeds smoothly, citizens will be satisfied with it. However, if environmental problems or a large-scale blowout or other type of accident occurs, probably the latent potential for citizen action against geothermal development will erupt. As one critic has noted, fervent supporters who become disenchanted can become formidable, knowledgeable, foes. Such use of public opinion has been reported recently in other states by foes of strip-mining and in the development of electric power plants in Montana, Wyoming, and Dakotas, Utah, Colorado, Arizona, and New Mexico.

If it is assumed that it is desirable to continue having citizens in favor of geothermal development, then (1) adequate notice of meetings and opportunities for involvement should be provided to all potentially affected persons, (2) opportunities for citizen input should be made available at all stages of the planning and development process.

The lack of understanding by the general public and the perceived need for information about geothermal development suggests that more formalized informational and educational programs need to be developed and made available. This recommendation assumes that an informed public and citizenry is less likely than an uninformed one to react negatively to geothermal production when large-scale development takes place in the future. Fostering citizen participation also probably would make the planning process one that is more actively controlling events, delineating viable options, maximizing future options and alternatives, and minimizing future uncertainty for the public, all of which should lead to a continued positive stance by citizens toward geothermal development.

The public opinion survey focused on public perception of possible environmental degradation rather than on public commitment to achieving environmental quality. Thus, in the future, there is a need to measure public preferences for allocating generated tax funds from geothermal development among various expenditure areas, including environmental quality. This is a shift away from concern about pregeothermal development as a source engendering possible social and environmental problems, toward a concern with what should be done with generated tax revenues after development has taken place.

While our research has stressed the probability that geothermal development will have relatively little population and employment impact upon Imperial County, in other areas that have undergone various kinds of development, the input of new economic resources has drastically influenced leadership patterns. Similarly, in Imperial County, the potential exists for large-scale impacts on the development of industry and population growth, including legal and illegal Mexican labor. In addition, if industry moved into Imperial County on a large-scale basis, it would bring a population with essentially different lifestyles from the existing one. Such population growth would trigger physical growth in towns and former rural areas requiring land use changes, additional services, and expanded community institutions such as in the political arena, education, religion, recreation, and so on. Thus, there needs to be at least some considered evaluation of the possible alternative growth patterns described in our analyses.

Finally, the large-scale development of geothermal resources in Imperial County probably will result in changed leadership and influence patterns if past research can be used as a reliable guide. The flow of such large-scale economic resources has to involve major decisions. There may be a severe disruption in what is now an extremely stable economic base. To what extent conflict will be engendered between the "old" agribusiness and "new" energy-related economic interests will be of utmost importance to the current leaders and citizens of Imperial County. The manner by which conflicts are ameliorated will influence the county for coming generations. More knowledge along these lines is essential if current leaders and citizens are to continue having a major impact in the county.

Research Conclusions on Public Opinion

The following seven key conclusions were drawn from the public opinion survey research presented in Chapter 6:

1. Almost 90% of the citizens are in favor of geothermal development.
2. Three-fourths believe geothermal development should be strictly regulated.
3. Three-fourths feel preserving the environment is important; $1/5$ saw environmental or social problems arising from geothermal development.
4. Expectations of geothermal development were as follows:
   (1) 20% expected cheaper energy
   (2) 20% expected greater power to Imperial County
   (3) 15% expected a variety of other positive benefits
   (4) 5% had negative expectations.
5. Those who did not care or were opposed to geothermal development had less than an average understanding of it.
6. Leaders were similar to citizens in favoring geothermal development showing environmental concerns and a desire for strict regulations.

Research Conclusions and Policy Recommendations 323

7. Issues of the past few years, such as water rights and geothermal leadership issues, have revealed a potential for conflict—devisiveness which could be brought into the public forum by a geothermal disaster, such as a blowout.

Only the last research finding on public opinion potential for conflict was excluded from the plan. Obviously, it is not wise for county leaders to suggest possible future conflict in a general plan.

Policy Recommendations from the Public Opinion Survey

The following six policy recommendations were made for the public opinion phase of the research:

1. Citizens should be allowed to be heard in the county planning process.
2. Adequate notice of meetings and opportunities for involvement should be given to all potentially affected persons.
3. Opportunities should be offered for citizen input at all stages of planning and development process.
4. Institutional and educational programs should be undertaken to develop an informed citizenry.
5. The potential for large-scale population change, differing life styles, and "boom towns" should be explicitly recognized.
6. Plans should be developed for the amelioration of potential conflicts between "old" agribusiness and "new" energy-related industries.

Policy recommendations from the public opinion study which were excluded from the General Plan fall into the following three categories:

Involvement of Citizens in Meetings, Planning, and Development for Geothermal Energy. This was only mentioned briefly in the research conclusion section of the Geothermal Element. It is not a research conclusion (citizens did not suggest this on questionnaires), but a policy recommendation of the researchers. It should be in the plan, otherwise there is only a *one-way* flow of information, from county to citizens, and this is clearly not beneficial over the long run.

Possible Large-Scale Population Change and Development of "Boom Towns." By favoring a population projection total of 232,000 persons in the year 2020, the county is implying that it expects more population. Therefore, it should examine the full consequences of such a figure. Again, for a long term "40-year" effect, as development unfolds, the current General Plan will probably be forgotten.

Reduction in Conflict between Old Agribusiness and New Geothermal Industry. This probably comes too close to home in the current agribusiness orientation of the county.

Recommendations

Assumptions

The recommendations presented below are based on five assumptions appearing generally applicable to western United States KGRAs. Since U.S. electrical geothermal development has to date only taken place at The Geysers in Northern California, these assumptions cannot actually be tested for Imperial County. At a future date, more specific information for Imperial County will undoubtedly become available to test those assumptions.

The first assumption is that a strong feeling of progeothermal development by both the local public and leadership is desirable for optimal development. Possible motivations for such a stance by local leaders early in the development process might include perception of future local business and tax revenues, future local industrialization, desire for personal advantage in ensuing leadership struggles, and desire to mitigate unfavorable developmental side effects which might be engendered by an apathetic leadership stance.

External leadership consists, among others, of heads of major private development firms, and important political and regulatory officials. In an interview study of private utility executives involved in geothermal development, Pivirotto (1976) listed the two most important pieces of information for prodevelopment decisions by executives to be knowledge of the resource and demonstration of environmental-technological feasibility in cases of low-quality geothermal resources, and knowledge of the resource and favorable economic analyses in cases of high-quality resources. Although internal and external leadership motivations would likely be quite different, and the public's motivation different again, an optimal development sequence rests on a strong proenergy stance from the start by all of these parties.

A second assumption is that encouragement of on-line electrical power at a very early stage in the development process would be a stimulus to development by filling gaps in research and knowledge. In reference to his experience as a county official in stimulating research on the geothermal resource in Imperial County, Pierson (1976) stated, "We are striving to avoid leaping to theoretical conclusions or

Research Conclusions and Policy Recommendations

conjectures based on inadequate observation of measurable data." Utility executives, when asked to choose from 18 possible government actions favorable to development, picked construction of a 10-MW$_e$ pilot plant as their second choice (after demonstration of well and brine flow). As noted earlier, Imperial County already has one 10-MW$_e$ pilot plant—Magma's, completed in 1979 (Figure 9-4), as well as the 10-MW SDG&E test loop facility completed in 1975 (Figure 9-5). As reported in Chapter 8, several more are in the planning and permitting stages.

A third assumption is that the extent of economic and technical feasibility of resource exploitation is of fundamental importance to any development situation. In regard to Imperial County, Biehler and Lee (1977) have pointed out a multitude of factors determining such feasibility, but perhaps the most critical are resource type (water-dominated, or hot dry rock), temperature, size of the resource, permeability and porosity, availability of adequate supply and quality of cooling water, and salinity of the geothermal brines. The better the resource on this multidimensional scale of characteristics, the more favorable the chances for successful development.

A fourth assumption is that the many types of federal, state,

FIGURE 9-4. Pipes at Magma Power pilot plant, East Mesa.

FIGURE 9-5. Pipes at San Diego Gas and Electric/Department of Energy pilot plant, Niland.

county, and local regulations will either favor or impede the development process. These regulations encompass leasing, exploratory drilling, discharge of wastes, power plant siting, occupational safety, power pricing, production, and many others. A recent study (Jet Propulsion Laboratory, 1976) estimated that in California, 9 federal, 16 state, and at least 3 local agencies play a role in geothermal development. This study also noted that at least nine years is required from the decision to explore to the production of on-line power. In addition to three levels of regulating authority, there are also three types of potentially developmental resources—federal, state, and private. Some of the legal aspects leading from a specific regulatory level to a specific land ownership type have not been worked out yet in the courts (an example is local jurisdiction over development on federal lands). For a regulatory pathway which is clearly established, there is considerable variation from state to state and locality to locality. Also, within one state, enforcement patterns may vary depending on local conditions and political factors.

A final assumption is that certain population-distance considerations may be important in determining the potential market for energy produced (Fowler, 1975). Because of the possibilities of transmitting energy hundreds of miles to distant markets, this consideration would appear most applicable to highly remote KGRAs. In the case of The Geysers, the remote, sparsely populated locale does not economically inhibit transmission into the northern California grid. If, however, large fields were discovered in northern Wyoming, market considerations would assume greater importance as a development consideration.

Inasmuch as each geothermal site is unique in some aspects, the study of each localized site will produce conclusions and recommendations specific to that area. The following discussion applies research conclusions to policy recommendations for Imperial County. The discussion is divided into recommendations tied most closely with the physical environment and socioeconomic factors. Transfer to other geothermal areas is possible in proportion to the similarities between them and Imperial County.

Environmental

(1) Inasmuch as no model is likely to satisfy the demands for knowledge about subsidence, seismicity, and injection, actual commercial-sized geothermal operations are needed. Active encouragement of pilot projects would expedite accumulation of the most meaningful data for a smooth development process.

(2) Experiments so far have identified no sure prospect for substantial environmental harm (though the possibility still exists). In this light, and combined with a high existing environmental quality and the need for operational data, small-scale development in farming areas on the order of one 50-MW_e plant for each discrete productive area would seem a worthwhile gamble. If, during such operations, the environment were placed under an undesirable burden which would be unacceptable at a large scale, continued operation at the small-scale for the sake of the original investment should not severely jeopardize the environment. An evaluation could then be made of future development. If, on the other hand, no appreciable environmental harm were registered, geothermal development could proceed with greater confidence and speed.

(3) Different areas of the county have different environmental sensitivities (e.g., crop differences, feed lot vs. crops, communities vs. fields), and the characteristics of geothermal fluids also vary greatly. For example, a spill of water from the Heber KGRA would have a far lesser impact than an outfall of fluids from the Salton Sea field which has ten times the dissolved solids and substantially higher temperatures. Hence, planning regulations and decisions should be justifiably concerned with intracounty, localized differences.

Likewise, it is appropriate to consider that each phase of development has attendant pollution and nuisance problems (e.g., dust from construction, noise from drilling, gases from operation) and that environmental sensitivities vary during construction, testing, and operation. In addition, crop sensitivity varies according to diurnal and seasonal patterns. To help relieve this problem, different phases of geothermal operation should be coordinated with crop peculiarities at each location.

Socioeconomic

(1) The general progeothermal feeling of Imperial County citizens should reduce the apprehension of county officials and politicians concerned with development effects. This favorable attitude is most strongly represented in older age categories. Certainly such a mandate for change is unusual if not rare. Such attitudes should expedite geothermal development, at least in the initial stages, and county officials may find that the relatively long delays usually characterizing energy developments will be shortened.

(2) Although the county labor pool is not presently trained for the construction and operation of geothermal power plants, it offers a valuable resource, especially for jobs auxiliary to geothermal power use. Initial geothermal programs will have little effect on workers, but as development continues and expands, high rates of unemployment and underemployment should decrease. A retraining program implemented well before skilled workers are needed would increase the potential benefits of geothermal development to local workers and thereby be helpful in encouraging young people to remain in the county.

(3) Creative regional geothermal planning would be beneficial. Present urban population borders two KGRAs and conflict between residents and development is possible. Through land use policy supported by enforcement measures, in-migrants might be led to locate

away from possible geothermal side effects such as noise, odors, and congestion. This would reduce the potential of incidents and speed resource development. Generally speaking, areas already explored with unfavorable results would appear to be the best candidates for new population centers. Of course, this procedure is also helpful because it gives later exploratory efforts a freer hand.

(4) An active exchange program of data and personnel between those who have worked on the development of geothermal energy south of Mexicali at Cerro Prieto (current capacity 150 MW$_e$) and the development fields of Imperial County would be mutually beneficial. This could be coordinated with other Mexican–U.S. agreements (e.g., water quantity and quality from the Colorado River, sewage discharge from Mexicali to the Salton Sea, and geotechnical collaboration between Mexican and U.S. geothermal developers).

(5) The monolithic economic structure of the county might be useful in encouraging the development of those aspects of the resource tending to be mutually beneficial for geothermal developers and agriculture, such as the use of nonelectric potential for food processing. Generally, companies approaching development and utilization from this perspective would likely encounter little objection and probably explicit encouragement.

(6) A broader but still stable economic base for the county should be encouraged. Industry will find a great variety of suppliers and markets in Imperial County, and in nearby counties. Recent research has indicated a significant population and employment growth in nonmetropolitan counties, particularly those such as Imperial, which are adjacent to metropolitan areas. Many of these counties have less than ideal human environments.

Large, adjacent labor pools and markets, established transportation and communication networks, snow-free climate, close proximity to coastal communities, open space, minimal air pollution, plus the availability of plentiful energy for nonelectric application could contribute toward making Imperial County a very attractive area for industrialization. Heating and cooling and perhaps some process heat could be provided at a cost substantially reduced from that of more conventional sources.

(7) Almost all geothermal electricity generated in the county will be exported. The right to buy varying portions of this electricity, or at least the right periodically to review and adjust the agreement, would gain the county added flexibility during periods of future growth.

Implications of Research-Oriented Planning to Imperial County and Other Counties

Imperial County took an unusual step for a rural county in commissioning a research project to form a major part of its General Plan. Many sophisticated urban counties would not go so far. An earlier section examined the actual use of the research results, based on the research model presented earlier, in the plan. In the population and public opinion areas, most of our research findings were utilized. However, few of our policy recommendations were used. Taken together, the use of research and recommendations indicates the research's success. Many other research inputs to governments have had virtually no effect on official plans.

One important reason for exclusion of some of the results may be our biases. However, all phases of our research were approved on content by both the county and an outside Industrial Advisory Committee. The other presumed reasons for exclusion include slow pace of occurrence, such as for population variables; sensitive issues, such as for geothermal unemployment; lack of sophistication for implementation, such as for use of regional data; overinterpretation, for the population projections; and desire for county control of citizen involvement.

Since planning is partly a political process, this may justify these deletions for the present time. The more important question is, will the deletions hurt the long-range plan in the changed county of, say, 1990 or 2020? In the future, the answer would appear to be yes, it will hurt. Clearly, the plan should be broadened. Possibly a new comprehensive research study could be commissioned in say, 15 years, to revamp or update the plan. Overall, however, we feel that the county's boldness in initiating the research, accepting the concept of a broad model, and using substantial research results is unusual and commendable.

Other counties faced with planning for geothermal development may also utilize these research results. Such counties will face the obvious problem that only a portion of the Imperial County research is transferable to their locality. However, integrated assessment of the model presented in Chapter 1 (Table 1–4) should assist transferability in its majority to other counties. Although political obstacles may stand in the way of practical utilization of a high percentage of model results, energy stringencies may lessen such obstacles in the near future.

It has been a major goal of this book to encourage local and state governments to adopt a broad model in order to plan creatively for regional geothermal energy utilization. The cost is reasonable. In an era of dwindling fossil fuel supplies, early recognition of the potential of alternative energy forms is essential.

Acknowledgment

Parts of this chapter were derived from an article published in *Energy* (1979). Dr. Martin J. Pasqualetti was a coauthor of that paper with us, and we owe him a great deal for helping develop the basic substance of that article.

Appendix: Public Opinion Questionnaire

UNIVERSITY OF CALIFORNIA, RIVERSIDE

BERKELEY • DAVIS • IRVINE • LOS ANGELES • RIVERSIDE • SAN DIEGO • SAN FRANCISCO SANTA BARBARA • SANTA CRUZ

COMMUNITY RESEARCH PROJECTS
3627A CANYON CREST DRIVE
RIVERSIDE, CALIFORNIA 92507

August 2, 1976

Dear Resident of Imperial County:

We are conducting research on potential geothermal energy development in Imperial County. It is necessary for proper planning that we find out how residents of Imperial County view geothermal development. This is an excellent opportunity for you to express your views and ideas about geothermal development.

Since it is impossible to ask every individual and family in Imperial County their view, we have carefully selected a sample of about 1400 households in the County and have sent them these questions. It is very important to you and to us that you complete these questions now and return them right away. This is the only way most people in Imperial County will have an opportunity to express their views on geothermal government. So that our report to the public truthfully reflects your views, we hope you will complete these questions. Please return them to us in the enclosed self-addressed, stamped envelope as soon as possible.

Your answers will be used in computer analysis. All findings will be made public. However, no individual, such as yourself, will be identified in reports.

Your help is necessary and we believe that your answers will help influence how geothermal development will take place in Imperial County.

Thank you for your help and involvement.

Sincerely,

Edgar W. Butler, Ph.D.
Principal Investigator

EWB/pn
Enclosures (2)

IMPERIAL COUNTY GEOTHERMAL PUBLIC SURVEY: 1976

WE APPRECIATE YOUR COOPERATION AND ARE INDEBTED TO YOU FOR ANSWERING THESE IMPORTANT QUESTIONS AND FOR ANY ADDITIONAL COMMENTS YOU WISH TO MAKE. THANK YOU.

QUESTIONNAIRE NO.

N⁰ 11478

PLEASE CIRCLE OR WRITE THE CORRECT ANSWER FOR YOU

107. WHAT IS YOUR UNDERSTANDING OF GEOTHERMAL DEVELOPMENT?
 1 very good
 2 fair
 3 average
 4 slight
 5 none

108. ARE YOU IN FAVOR OR OPPOSED TO GEOTHERMAL DEVELOPMENT IN IMPERIAL VALLEY?
 1 strongly in favor
 2 in favor
 3 don't care
 4 opposed
 5 strongly opposed

109. DO YOU FEEL MOST PEOPLE IN IMPERIAL VALLEY ARE IN FAVOR OR OPPOSED TO GEOTHERMAL DEVELOPMENT?
 1 strongly in favor
 2 in favor
 3 don't care
 4 opposed
 5 strongly opposed

110. DO YOU THINK MOST CIVIC LEADERS IN IMPERIAL VALLEY ARE IN FAVOR OR OPPOSED TO GEOTHERMAL DEVELOPMENT?
 1 strongly in favor
 2 in favor
 3 don't care
 4 opposed
 5 strongly opposed

111. DO YOU AGREE OR DISAGREE THAT GEOTHERMAL DEVELOPMENT SHOULD BE STRICTLY REGULATED?
 1 strongly agree
 2 agree
 3 uncertain
 4 disagree
 5 strongly disagree

WHY? _____

112. IMPERIAL COUNTY GEOTHERMAL DEVELOPMENT SHOULD BE
 1 prohibited
 2 closely regulated
 3 un-regulated

113. IF GEOTHERMAL DEVELOPMENT IS COMPLETED, PROPERTY VALUES WILL:
 1 greatly increase
 2 increase somewhat
 3 remain the same
 4 decrease somewhat
 5 greatly decrease

114. WHAT DO YOU EXPECT FROM GEOTHERMAL DEVELOPMENT?

Public Opinion Questionnaire

115. PRESERVING THE ENVIRONMENT IS AN IMPORTANT ISSUE IN GEOTHERMAL DEVELOPMENT?
    1 yes
    2 unsure
    3 no

116. DO YOU FORSEE ANY ENVIRONMENTAL OR SOCIAL PROBLEMS ARISING OUT OF GEOTHERMAL DEVELOPMENT?
    1 yes
    2 unsure
    3 no

117. IF YES: WHAT KIND OF PROBLEMS?
    _____
    _____
    _____
    _____

118. ARE YOU AWARE OF ANY OPPOSITION IN IMPERIAL COUNTY TO GEOTHERMAL DEVELOPMENT?
    1 yes
    2 unsure
    3 no

119. IF YES: WHO ARE THE INDIVIDUALS AND/OR GROUPS AGAINST IT?
    _____
    _____
    _____
        1 no opposition known

120. IF YES: WHY ARE THEY OPPOSING IT?
    _____
    _____
    _____
        1 no opposition known

121. TO YOUR KNOWLEDGE, HAS THERE BEEN A MEETING HELD TO EXPLAIN OR DISCUSS GEOTHERMAL DEVELOPMENT?
    1 yes
    2 no
    3 don't know

122. HAVE YOU EVER ATTENDED A MEETING ON POTENTIAL GEOTHERMAL DEVELOPMENT?
    1 yes
    2 no

123. IF YES, HOW DID YOU LEARN ABOUT SUCH A MEETING?
    1 direct mail
    2 newspaper
    3 friend or neighbor
    4 other (please specify): _____
    5 not attended

124. DO YOU FEEL THAT CITIZENS OF IMPERIAL VALLEY HAVE BEEN ADEQUATELY INFORMED ABOUT GEOTHERMAL DEVELOPMENT?
    1 yes - WHO? _____
    2 no
    3 unsure

125. OUTSIDE OF THIS QUESTIONNAIRE, HAS ANYONE ASKED YOU HOW YOU FEEL ABOUT GEOTHERMAL DEVELOPMENT?
    1 yes - WHO? _____
    2 no
    3 unsure

126. HAVE YOU EVER VISITED A GEOTHERMAL TESTING SITE?
    1 yes
    2 no

Do you agree or disagree with the following statements? Please circle your answer.

1 strongly agree
2 agree
3 uncertain
4 disagree
5 strongly disagree

#					Statement
127.	1	2	3	4	5	Geothermal development will bring new tax revenues to Imperial County.
128.	1	2	3	4	5	Noise from geothermal development can be bothersome.
129.	1	2	3	4	5	Economic benefits from geothermal development are more important than environmental costs.
130.	1	2	3	4	5	Because it will attract new residents, I'm against geothermal development.
131.	1	2	3	4	5	The construction of geothermal power plants, transmission lines, pipelines and roads which result from geothermal development will create eyesores.
132.	1	2	3	4	5	Because it will attract new businesses and help Imperial Valley grow, I'm in favor of geothermal development.
133.	1	2	3	4	5	Most geothermal electricity produced in Imperial County should be used in Imperial County.
134.	1	2	3	4	5	A fuel shortage will develop in the United States unless geothermal and other sources of energy are developed.
135.	1	2	3	4	5	Geothermal energy will provide cheap electricity for Imperial Valley.
136.	1	2	3	4	5	I like Imperial Valley the way it is, and don't want it to change.
137.	1	2	3	4	5	New developments like geothermal are not welcome in Imperial County.
138.	1	2	3	4	5	Most geothermal electricity produced in Imperial County will be used in Imperial County.
139.	1	2	3	4	5	Imperial County can broaden its economic emphasis to more than agriculture through geothermal development.
140.	1	2	3	4	5	Geothermal companies should have the main responsibility to plan and conduct steam exploration and production.
141.	1	2	3	4	5	Geothermal development may cause unusual odor problems.
142.	1	2	3	4	5	Geothermal development will increase demands on city and county government and thus increase taxes.
143.	1	2	3	4	5	Geothermal development will increase jobs in Imperial County.
144.	1	2	3	4	5	Local government officials have primary responsibility to plan geothermal exploration and production.
145.	1	2	3	4	5	Geothermal development will take water away from agriculture.
146.	1	2	3	4	5	Geothermal resources in Imperial Valley should be used for purposes other than electricity, such as by industry or for chemicals.
147.	1	2	3	4	5	Geothermal development will result in fewer Mexican National agricultural workers crossing daily into Imperial Valley.
148.	1	2	3	4	5	The Imperial County policy that new industries, like geothermal, should be able to live with agriculture is a good one.
149.	1	2	3	4	5	Geothermal development will cause border regulations to change making it easier for Mexican National workers to cross into the United States.

Public Opinion Questionnaire

150. WHERE DID YOU LIVE BEFORE COMING TO IMPERIAL VALLEY?

    _____ _____
    (city or town)   (state)
        1   lived entire life in Imperial County

151. HOW LONG HAVE YOU LIVED IN IMPERIAL VALLEY?
    1. entire life
    ___ years

152. HOW LONG HAVE YOU LIVED IN THIS HOUSE?
    1. less than six months
    2. 6 months to two years
    3. two to five years
    4. five to ten years
    5. over ten years

153. DO YOU PLAN TO LIVE IN IMPERIAL VALLEY THE REST OF YOUR LIFE?
    1. yes
    2. uncertain
    3. no

154. WHAT WOULD MAKE IMPERIAL VALLEY A BETTER PLACE TO LIVE?
    _____
    _____
    _____

156. HOW MUCH RURAL LAND DO YOU OWN IN IMPERIAL VALLEY?
    0. none   4. 40-99 acres
    1. 1-5 acres   5. 100-319 acres
    2. 6-10 acres   6. 320-500 acres
    3. 11-39 acres   7. 500 or more acres

157. HOW MUCH LAND DO YOU OWN WITHIN CITY LIMITS?
    0. none
    1. 1-5 acres
    2. 6-10 acres
    3. 10 or more acres

158. DO YOU LIVE IN A:
    1. single family residence
    2. duplex
    3. apartment
    4. condominium
    5. mobile home or trailer
    6. other

159. DO YOU:
    1. own
    2. rent
    3. other

160. HOW MANY YEARS OF SCHOOL HAVE YOU COMPLETED?
    ___ years

161. ABOUT HOW MANY MONTHS DID YOU WORK DURING 1975, FROM JAN. 1 to DEC. 31, 1975?
    1. less than 3 months
    2. 3 months to 5 months
    3. 6 months to 11 months
    4. 12 months, counting vacation time

162. WHAT KIND OF JOB DO YOU HAVE? WHAT DO YOU DO ON THAT JOB?
    _____
    _____
    _____
    00   no job

164. HOW MANY PEOPLE ARE IN THIS HOUSEHOLD? THAT IS, ALL PERSONS WHO MAKE THIS THEIR PERMANENT ADDRESS? (Be sure to include infants under one year of age).
    _____

165. HOW MANY PEOPLE IN THIS HOUSEHOLD ARE CITIZENS OF THE UNITED STATES?
    _____

166. WHAT WAS THE TOTAL INCOME IN 1975 OF ALL THE PERSONS WHO LIVE IN THIS HOUSEHOLD?
    1. under 2,999   4. 9,000 - 14,999
    2. 3,000 - 4,999   5. 15,000 - 24,999
    3. 5,000 - 8,999   6. 25,000 +

167. PLEASE INDICATE
    1. Anglo
    2. Mexican-American or Chicano
    3. Mexicano or Citizen of Mexico
    4. Black or Negro
    5. Asian-American
    6. American Indian
    7. Other _____

ANY COMMENTS YOU WOULD LIKE TO MAKE?
_____
_____
_____
_____
_____
_____

THANK YOU FOR YOUR COOPERATION

References

Allen, W., and H. K. McCluer, "Abatement of Hydrogen Sulfide Emissions From The Geysers Geothermal Power Plant," *Proceedings of the 2nd United Nations Symposium of the Development and Use of Geothermal Resources,* Washington, D.C.: U.S. Government Printing Office, pp. 1313–1315, 1976.

Anderson, David N., *Recent Developments of Geothermal Resources in the United States.* Paper presented at the Arid Lands Association, Lake Tahoe, Nevada, April, 1979.

Anderson, David N. *Recent Developments of Geothermal Resources in the United States.* Paper presented at the Geothermal Resources Council, Special Short Course No. 6, December 12–13, Houston, Texas, 1977.

Armstead, Christopher, "Geothermal Economics," in *Geothermal Energy.* Geneva: UNESCO, 1973.

Armstrong, DeVonne W., and Edgar W. Butler, *Eastside Community Plan.* Riverside, California: City of Riverside, January, 1974.

Axtmann, R. C., "Environmental Impact of a Geothermal Power Plant," *Science,* **187**(4179): 795–803, 1975.

Baer, Roger K., "Male Labor Force Participation Revisited," *Demography* **9**(4):635–654, 1972.

Baker, Joseph N., "A City Invests in Its Future," in *Proceedings, Conference on Research for the Development of Geothermal Energy Resources.* National Science Foundation, pp. 335–339: 1974.

Barnea, Joseph, "Geothermal Power," *Sci. Am.* **226**(1): 70–77, 1972.

Beale, Calvin L., *The Revival of Population Growth in Nonmetropolitan America.* Washington: Department of Agriculture, Economic Research Service, Pub. No. ERS-605, 1975.

Beale, Calvin L., "A Further Look at Nonmetropolitan Population Growth Since 1970," *Am. J. Agric. Econ.* **50**(5): 953–958, 1976.

Biehler, Shawn, and Tien Lee, *Final Report on a Resource Assessment of the Imperial Valley.* Riverside: University of California, Dry Lands Research Institute, 1977.

Biggar, J. C., and E. W. Butler, "Fertility and its Interrelationships with Population Size in a Southern State," *Rural Sociol.* **34**:528–536, 1969.

Birsic, Rudolph, *The Geothermal Steam Story.* Fullerton: R. J. Birsic, 1974.

Bogue, Donald J., *The Population of the United States.* Glencoe (Illinois): The Free Press, 1959.

Boughey, Arthur S., D. Wilkin, and J. B. Pick, "Hydrology and Water Use in Kern County: A Modelling Study of Human Carrying Capacity." Unpublished paper presented at the Eighth American Water Resources Conference, St. Louis, 1972.

Bradshaw, Benjamin S., "Potential Labor Force Supply, Replacement, and Migration of Mexican American and Other Males in the Texas-Mexico Border Region," *Int. Migration Rev.* **10**(1): 26–29, 1976.
Bradshaw, Benjamin S., and Frank Bean, "Trends in the Fertility of Mexican Americans: 1950–1970," *Social Science Quarterly* (March, 1973):688–696.
Buck J. Vincent, *Regulatory Planning and Policy Aspects of Geothermal Energy Development in Imperial County, California*. Final Report on NSF/ERDA Grant No. AER 75-08793, 1977.
Bufe, C. G., and F. W. Lester, "Seismicity of The Geysers—Clear Lake Region, California," *EOS Am. Geophysical Union Transaction*, **56**:1021, 1975.
Butler, Edgar W., and James B. Pick, "Evaluation Assessments for the General Plan of Imperial County, California: Some Practical Implications." Paper presented at the Annual URISA Conference, Washington, D.C., August 9, 1978.
Butler, Edgar W., and James B. Pick, *Opinion About Geothermal Development in Imperial County, California, Final Report*, Dry Lands Research Institute, Riverside, California, DLRI Report No. 17, 1977.
Butler, Edgar W., *The Urban Crisis: Problems and Prospects in America*. Santa Monica, California: Goodyear Publishing Company, 1977.
Butler, Edgar W., *Urban Sociology*. New York: Harper and Row, 1976.
Butler, Edgar W., F. Stuart Chapin, George C. Hemmens, Edward J. Kaiser, Michael A. Stegman, and Shirley F. Weiss. *Moving Behavior and Residential Choice: A National Survey*. National Cooperative Highway Research Program Report 81, Highway Research Board. Washington, D.C.: National Academy of Sciences, 1969.
California State Legislature, Assembly Bill No. 2644. Sacramento, California State Legislature, 1978.
Castrantas, Harry M., Thomas A. Turner, and Robert W. Rex, "Hydrogen Sulfide Abatement in Geothermal Steam," in *Geothermal Environmental Seminar—1976*, Fayne L. Tucker and Mary Dean Anderson, (eds.), Shearer Graphic Arts, Lakeport, California, pp. 139–159, 1976.
Cataldi, R., P. DiMario, and T. Leardini, "Application of Geothermal Energy to the Supply of Electricity in Rural Areas," *Geothermics* **2**(1): 3–16, 1973.
Chasteen, A.J., "Geothermal Steam Condensate Reinjection," in *Proceedings of the Conference on Research for the Development of Geothermal Energy*, Pasadena, California, pp. 340–344, 1974.
City of El Centro, *City of El Centro Energy Utility Core Field Experiment*. Vol. 1. Technical Proposal. El Centro, California: City of El Centro, July 14, 1978.
Collver, O. A., "Women's Work Participation and Fertility in Metropolitan Areas: An International Comparison," *Demography* **5**:55–60, 1968.
Collver, O. A., and E. Langlois, "The Female Labor Force in Metropolitan Areas: An International Comparison," *Economic Development Cultural Change* **10**(4): 367–385, 1962.
Comptroller General, *Public Involvement in Planning Public Works Projects Should Be Increased*. Comptroller General of the United States, 1974.
Dagodog, W. Tim, "Source, Regions and Composition of Illegal Mexican Immigration to California," *International Migration Review* **9**(4): 490–512, 1975.
Dambly, Benjamin W., "Heat Exchanger Design for Geothermal Power Plants," *Proceedings of the 13th Intersociety Energy Conversion Engineering Conference*. Warrendale (Pennsylvania), Society of Automotive Engineers, pp. 1102–1108: 1978.
Davis, Cal, Preliminary Scenario, Imperial Valley Region, State of California, unpublished report, Jet Propulsion Laboratory, Pasadena, April 19, 1976.

References

Defferding, Leo J., and Ron A. Walter, "Disposal of Liquid Effluents from Geothermal Installations," *Transactions of the Geothermal Resources Council*, **2**:141–144, 1978.

DeSandre, P., "Recherche par micro-zones sur les relations entre fecondite et certaines characteristiques du milieu economique et social dans une region italienne," *Genus* **27**: 1–27, 1971.

Dibblee, T. W., Jr., "Geology of the Imperial Valley Region, California," *Geology of Southern California*, California Division of Mines, Bulletin No. 170, pp. 20–28, 1954.

Dixon, W. J. (ed.), *Biomedical Computer Programs (BMDP)*. Berkeley: University of California Press, 1976.

Draper, N. R., and H. Smith, *Applied Regression Analysis*. New York: Wiley, 1966.

Dunlap, Riley E., and Don A. Dillman, "Decline in Public Support for Environmental Protections: Evidence from a 1970–1974 Panel Study. *Rural Sociology* **41**(Fall, 1976): 382–390.

Easterlin, R. A., "The American Baby Boom in Historical Perspective," *American Economic Review* **51**: 869–911, 1961.

Edmunds, Stahrl, and Adam Z. Rose (eds.), *Geothermal Energy and Regional Development*. New York: Prager, 1979.

Eisenbus, Robert A., and Robert B. Avery, *Discriminant Analysis and Classification Procedure*. Lexington, Massachusetts: D.C. Heath and Co., 1972.

Eisenstat, Samuel M., "Federal Tax Treatment of Geothermal Exploration and Production," *Proceedings of a Conference on the Commercialization of Geothermal Resources*. Davis, California: Geothermal Resources Council, pp. 61–64, 1978.

El Centro Chamber of Commerce, *Community Economic Profile*. El Centro: El Centro Chamber of Commerce, 1976.

Elders, W. A., "Regional Geology of the Salton Trough," *Geothermal Development of the Salton Trough, California and Mexico*, (T. D. Palmer et al. Eds.), Lawrence Livermore Laboratory, University of California, pp. 1–12, 1975.

Elders, W. A., R. W. Rex, T. Meidav, P. T. Robinson, and S. Biehler, "Crustal Spreading in Southern California," *Science* **178**: 15–24, 1972.

Employment Development Department, State of California, *Annual Planning Information, Imperial County, California*. Los Angeles, Employment Development Department, 1978.

ERDA-DGE Postulated Development Scenario (MW_e), and Scenario Summary, Washington, D.C. March, 1977.

Environmental Systems Research Institute, *Automap II, User's Manual*. Redlands, California: Environmental Systems Research Institute, 1975.

Ermak, Donald L., "A Scenario for Geothermal Electric Power Development in Imperial Valley," *Energy* **3**: 203–217, 1978.

Evans, D. M., "Man-Made Earthquakes in Denver," *Geotimes* **10**(9): 11–18, 1966.

Finnemore, E. John, "The Analysis of Subsidence Associated with Geothermal Development and its Potential for Environmental Impact: A Summary Description of the Present Status of Geothermal Reservoir and Subsidence Models," Technical Memorandum No. 5139-400-09, Systems Control, Inc., Palo Alto, 1976.

Force, Peter R., Adaptation of Input–Output Technique to a Small Economic Area. Master's thesis. Los Angeles: California State College, 1970.

Fowler, John M., *Energy and Environment*. New York: McGraw-Hill, 1975.

Freudenberg, William R., "A Social Impact Analysis of a Rocky Mountain Energy Boomtown." Paper presented at American Sociological Association annual meetings, San Francisco, 1977.

Fulton, Maurice, "Industry's Viewpoint of Rural Areas," in Whiting, Larry R. (ed.), *Rural Industrialization*. Ames: Iowa University Press, 1974.

Galle, Omer R., John D. McCarthy, and Walter R. Gove, "Overcrowding and Isolation: Some Behavioral Consequences." Paper presented at annual meeting of Population Association of America, New York City, April 18-20, 1974.

Galle, Omer R., John D. McCarthy, Walter R. Gove, and William Zimmerin, "Population Density, Social Structure, and Interpersonal Violence: An Intermetropolitan Test of Competing Models." Paper presented at American Psychological Association Meetings, Montreal, Canada, August 29, 1973.

Gelbner, Christopher G., "Regional Differences in Employment and Unemployment, 1957-72," *Monthly Labor Review* **97**(3): 15-24, 1974.

Geothermal Hot Line, "Department of Energy Plans Power Project." *Geothermal Hot Line* **8**(3): 7 October, 1978.

Geothermal Hot Line, "New Geothermal Project Planned in Imperial County." *Geothermal Hot Line* **8**(2): 3 July, 1978.

Geothermal Resources Council, *Final Report to the Department of Energy by the Geothermal Resources Council on an Industry Survey of the Need for a Federally Funded Geothermal Demonstration Plant*. Davis, California: Geothermal Resources Council, 1978.

Geothermal Resources Council, KGRA Descriptions for Imperial Valley Field Trip, December 1, 1978.

The Geyser, "Magma Electric's-East Mesa Plant," *The Geysers*, **5**(6): 1-2, November 17, 1978.

Ginther, Jerry D., Howard E. Lindow, G. A. Hornberger, and Robert W. Shively, "Corporate and Community Decision Making for Locating Industry," in Whiting, Larry R. (ed.), *Rural Industrialization*. Ames: Iowa University Press, 1974.

Goldberg, Kalman, and Robin Linstromberg, "A Revision of Some Theories of Economic Power," *Quarterly Review of Economics and Business*, **6**(Spring, 1966): 7-17.

Goldman, G., and D. Strong, *Governmental Costs and Revenues Associated with Geothermal Energy Development in Imperial County*. Lawrence Livermore Laboratory, Livermore, California, UCRL-13800, 1977.

Goldsmith, Martin, "Engineering Aspects of Geothermal Development in the Imperial Valley," EQL Memorandum No. 20. Pasadena, California. Environmental Quality Laboratory, California Institute of Technology, 1976a.

Goldsmith, Martin, "Geothermal Development and the Salton Sea," EQL Memorandum No. 17. Pasadena, California. Environmental Quality Laboratory, California Institute of Technology, 1976b.

Gray, Irwin, "Employment Effects of a New Industry in a Rural Area," *Monthly Labor Review* **91**(6): 26-30, 1969.

Gregler, Leo, Joan W. Moore, and Ralph G. Gusman, *The Mexican-American People*. New York: The Free Press, 1970.

Green, Phyllis, and Maureen Farnan, "Regulatory Policy, Political Participation, and Social Implications of Geothermal Resource Development in the Imperial Valley," Riverside: University of California: Center for Social and Behavioral Science Research, January, 1977.

Greenwood, Michael J., and Patrick J. Gormely, "A Comparison of the Determinants of White and Nonwhite Interstate Migration," *Demography* **8**(1): 141-155, 1971.

Greenwood, Michael J., and Douglas Sweetwood, "The Determinants of Migration Between Standard Metropolitan Statistical Areas," *Demography* **9**(4): 665-681, 1972.

References

Griscom, T., and L. J. P. Muffler, *Salton Sea Aeromagnetic Maps*. Washington, D.C.: U.S. Geological Survey, Map GP-754, 1971.

Handler, Philip, and Judith Sherwood, "The Plato System Population Dynamics Course," in Greville, T. N. E. (ed.), *Population Dynamics*, New York: Academic Press, 1972.

Harbon, Don C., "Financing of a Commercial Geothermal Development: An Operators Objective and Constraints," *Proceedings of a Conference on the Commercialization of Geothermal Resources*. Davis, California: Geothermal Resources Council, 1978.

Haren, Claud, "Location of Industrial Production and Distribution," in Whiting, Larry R. (ed.), *Rural Industrialization*. Ames: Iowa University Press, 1974.

Hauser, Philip, "The Census of 1970," *Scientific American* **225**(1): 17–25, 1971.

Hawley, Amos H., *Urban Society*, New York: Ronald Press, 1971.

Heer, D. M., and J. W. Boynton, "A Multivariate Analysis of Differences in Fertility in the United States Counties," *Social Biology* **17**:180–194, 1970.

Heer, D. M., and E. S. Turner, "Areal Differences in Latin American Fertility," *Population Studies* **18**:279–292, 1965.

Hernandez, Jose, Leo Estrada, and David Alvirez, "Census Data and the Problem of Conceptually Defining the Mexican-American Population," *Social Science Quarterly* (March, 1973): 671–687.

Hernandez, Juan P., "El Comercio, La Zona Libre y Algo mas en el Municipio de Mexicali," *Boletin Economico* **3**(1): 5–25, 1974.

Hill, D. P., P. Mowinchel, and L. G. Peake, "Earthquakes, Active Faults, and Geothermal Areas in the Imperial Valley, California," *Science* **188**: 1306–1308, 1975.

Hinrichs, Thomas C., and Harry W. Falk, Jr., "The East Mesa Magmamax Process Power Generation Plant." *Transactions of the Geothermal Resources Council* **1**: 141–142, 1977.

Holt, Ben, and John Brugram, "Investment and Operating Costs of Binary Cycle Geothermal Power Plants," in *Proceedings, Conference on Research for the Development of Geothermal Energy Resources*. National Science Foundation, pp. 292–300, 1974.

Hunter, L. C., G. L. Reid, and D. Boddy, *Labor Problems and Technological Change*. London: George Allen and Irwin Ltd., 1970.

Imperial County, *Geothermal Element*. El Centro, County of Imperial, 1977.

Imperial County, *Imperial County Goals*. El Centro, County of Imperial, 1975.

Imperial County, *Ultimate Land Use Plan*. El Centro, County of Imperial, 1973.

Imperial County, *Conservation Element*. El Centro, County of Imperial, 1973.

Imperial County, *Overall Economic Development Plan*. El Centro: Imperial County Economic Development Commission, 1970.

Imperial Valley Action Plan, Utility Technical Committee, *Transmission Planning Corridors for 2,000 MW Geothermal Development*. El Centro, California Imperial Valley Action Plan, October, 1978.

"Information Presented at the Geothermal Project Planning Council, Menlo Park, California, June 21–23, 1977." MITRE Working Paper 12841, McLean, Virginia, July 15, 1977.

James, R., "The Economics of the Small Geothermal Power Station," *Geothermics Special Issue*, **2**:1697–1704, 1970.

Jessen, Raymond J., *Statistical Survey Techniques*. New York: Wiley-Interscience, 1978.

Jet Propulsion Laboratory, *Geothermal Energy in California*. JPL Document No. 5040-25, Revision A, June, 1976.

Jet Propulsion Laboratory, *Report on the Status of Geothermal Energy Resources in California*, 5040-25. Pasadena, California: California Institute of Technology, March 31, 1976.

Johnson, C., *Effects of Geothermal Development on the Agricultural Resources of the Imperial Valley*. Riverside: University of California, Dry Lands Research Institute, 1977.

Johnson, Claude, *et al.*, Agricultural Resources. Preliminary Table, NSF Imperial County Geothermal Project, 1976.

Johnston, David, "Foes of Strip-Mining Learn How to Use Public Opinion," *Los Angeles Times*, 1976.

Jones, Bernie, and Charles Cortese, "Patterns of Boom Town Experiences: Implications For Future Work in the Field of Social Impact Assessment." Paper presented at the Society for the Study of Social Problems, New York City, August 27–30, 1976.

Kelly, R. E., "Atmospheric Dispersion and Noise Propagation at the Imperial Valley Geothermal Fields." Mimeo, draft, Lawrence Livermore Laboratory, University of California, Livermore, California, 1976.

Keyfitz, Nathan, and Wilhelm Flieger, *Population: Facts and Methods of Demography*. San Francisco: W. H. Freeman, 1971.

Kjos, Kaare Saxe, "The Role of Industry in the Socio-Economic Development of Calexico." Master's thesis, United States International University, 1974.

Kuhn, Philip H., "The Implications of Interdivisional Migration For Future Energy Consumption." Paper presented at the annual meeting of the Population Association of America, St. Louis, 1977.

Lachenbruch, Peter, *Discriminant Analysis*. New York: Hafner Press, 1975.

Lansing, John B., and Eva Mueller, *The Geographic Mobility of Labor*. Ann Arbor: Survey Research Center, 1967.

Lapp, Ralph, *The Logarithmic Century*. Englewood Cliffs, New Jersey: Prentice-Hall, 1973.

Larson, Tod, *The Costs of Geothermal Energy Development*. Riverside, California: Dry Lands Research Institute, University of California, 1977.

Lasker, R., R. H. Tenaza, and L. L. Chamberlain, "The Response of Salton Sea Fish Eggs and Larvae to Salinity Stress," *California Fish and Game* **58**(1): 58–66, 1972.

Lasswell, Harold D., and Abraham Kaplan, *Power and Society*, New Haven, Connecticut: Yale University Press, 1950.

Layton, D. W. (ed.), *A Technology Assessment of Geothermal Development in the Imperial Valley of California*, Vol. II. Lawrence Livermore Laboratory, Livermore, California, 1979.

Layton, D. W., *Water for Long Term Geothermal Energy Production in the Imperial Valley*. Lawrence Livermore Laboratory, Livermore, California, 1978.

Layton, D. W. (ed.), *A Technology Assessment of Geothermal Development in the Imperial Valley of California*, Vol. 2, Lawrence Livermore Laboratory, Livermore, California, 1979.

Layton, D. W., and David Ermak, "A Description of Imperial Valley, California for the Assessment of Impacts of Geothermal Energy Development." UCRL-52121, Lawrence Livermore Laboratory, 1976.

Lee, Tien-Chang, "Earthquake Potential and Geothermal Energy Extraction." Prepared under NSF/ERDA Grant No. AER 75-08793, NTIS, Virginia, 1977.

Leigh, J. A., A. Cohen, W. Jacobsen, and R. Trehan, *Site-Specific Analysis of Geothermal Development*. Vol. 1, Summary Report, MITRE Corporation (MTR-786), Washington, D.C., U.S. Department of Energy, August, 1978.

Leitner, P., An Environmental Overview of Geothermal Development: The Geysers—Calistoga KGRA. UCRL-52496. Livermore, California, Lawrence Livermore Laboratory, 1978.

Lessing, Lawrence, "Power From the Earth's Own Heat," in Boughey, A. S. (ed.), *Readings in Man, The Environment of Human Ecology*. New York: Macmillan, 1973.

References

Levine, Louis, "Unemployment by Locality and Industry," in *The Measurement and Behavior of Unemployment*. Princeton: Princeton University Press, 1957.

Lienau, Paul J., and John W. Lund (ed.), *Multipurpose Use of Geothermal Energy*. Klamath Falls: Oregon Institute of Technology, 1974.

Lienau, Paul J., "Space Conditioning with Geothermal Energy." *Proceedings of a Conference on the Commercialization of Geothermal Resources*, Davis, California, Geothermal Resources Council, pp. 23–26, 1978.

Lindal, B., "The Production of Chemicals From Brine and Seawater Using Geothermal Energy," *Geothermics*, Special Issue 2, 1970.

Lindal, B., "Industrial and Other Applications of Geothermal Energy," in Armstead, H.C.H., *Geothermal Energy*. Paris: UNESCO Press, 1973.

Lindsay, Donald R., "Procedural Requirements for Developing Geothermal Resources From Leasing to Power Generation." *Proceedings of a Conference on the Commercialization of Geothermal Resources*. Davis, California: Geothermal Resources Council, 1978.

Linton, A. M., "Innovative Geothermal Uses in Agriculture," in Lienau, Paul J., and John W. Lund (eds.), *Multipurpose Use of Geothermal Energy*. Klamath Falls: Oregon Institute of Technology, 1974.

Lofgren, B. E., "Measuring Ground Movement in Geothermal Areas of the Imperial Valley, California," in *Proceedings of the Conference on Research for the Development of Geothermal Energy Resources*, Pasadena, California, 1974.

Lofting, Everard, "The Regional Economy," in Edmunds, Stahrl, and Adam Rose (eds.), *Geothermal Energy and Resources Development*, New York: Praeger, 1979.

Lofting, Everard, *A Multisector Analysis of the Imperial County Economy*, University of California, Riverside: Dry Lands Research Institute, 1977.

Lofting, Everard, and Adam Rose, Economic Analysis for NSF-Imperial County Geothermal Project. Preliminary report on input-output analysis, 1976.

Magma Power Company, 1975 Annual Report. Los Angeles Magma Power Company.

May, R. C., "Effects of Salton Sea Water on the Eggs and Larvae of *Bairdiella icistia*.." *California Fish and Game*, 62(2): 119–131, 1976.

Meidav, Mae Z., and H. Tsvi Meidav, "Impact of Geothermal Energy Development in the Heber Area of Imperial Valley, California," *Transactions of the Geothermal Resources Council*, 1:217–219 (1977).

Meidav, Mae Z., "Methodology of Socioeconomic Assessments of Geothermal Energy Development," *Transactions of the Geothermal Resources Council*, 2(1):433–437 (1978).

Meidav, Tsvi, S. Sanyal, and Giancarlo Facca, "A Short Update on Worldwide Geothermal Activity," *Geothermal Energy Magazine* 5(5): 30–34, 1977.

Meidav, Tsvi, "Overview of Worldwide Geothermal Developments." *Proceedings of a Conference on the Commercialization of Geothermal Resources*. Davis, California, Geothermal Resources Council, pp. 7–10, 1978.

Melvin-Howe, G., "The Geography of Death," *New Scientist* 40: 612–614, 1968.

Morrison, Peter A., "Chronic Movers and the Future Redistribution of Population: A Longitudinal Analysis," *Demography* 8(2): 171–184, 1971.

Morrison, Peter A., Population Movements and the Shape of Urban Growth. Santa Monica: RAND, 1972.

Muffler, L. J. P. (ed.), *Assessment of Geothermal Resources of the United States—1978*. Menlo Park, California, U.S. Geological Survey, 1979.

Myers, George C., "Ecological Perspectives on Energy." Paper presented at the annual meeting of the Population Association of America, St. Louis, 1977.

Nakamura, S., "Economics of Geothermal Electric Power Generation at Matsukawa," *Geothermics* Special Issue 2, 1970.

Nathan Associates. *Industrial and Employment Potential of the United States—Mexico Border.* Washington: U.S. Department of Commerce, Economic Development Administration, 1968.

National Petroleum Council, *Environmental Conservation. The Oil and Gas Industries,* A National Petroleum Council Study, Vol. 1, 1971.

Otte, Carl, "Geothermal Energy Comes of Age," *Geothermal Energy Magazine,* **6**(8): 32–36, 1978.

Otte, Carl, "Drilling Production and Disposal Technology in the Salton Sea Area, Imperial County, California," in *Compendium of Papers—Geothermal Hearings,* The Resource Agency, Sacramento, California, 1970.

Panawek, Gregory, "Leveraged Leasing of a Geothermal Power Plant." *Proceedings of a Conference on the Commercialization of Geothermal Resources.* Davis, California: Geothermal Resources Council, pp. 55–58, 1978.

Pasqualetti, Martin J., James B. Pick, and Edgar W. Butler, "Geothermal Energy in Imperial County, California: Environmental, Socio-Economic, Demographic, and Public Opinion Research Conclusions and Policy Recommendations," *Energy* **4**: 67–80, 1979.

Paulson, Wayne, Edgar W. Butler, and Hallowell Pope, "Community Power and Public Welfare," *American Journal of Economics and Sociology* **28**(January, 1969): 17–27.

Pearson, Russ, "Uses of Geothermal Heat for Sugar Refining," in *Direct Utilization of Geothermal Energy: A Symposium.* Davis, California: Geothermal Resources Council, pp. 119–120, 1978.

Penner, S. S., and L. Icerman, *Energy,* Vol. 1. Reading, Massachusetts: Addison-Wesley Publishing Company, 1974.

Perry, G., *Unemployment Flows in the U.S. Labor Market. Brookings Paper on Economic Activity* No. 2: 245–292, 1972.

Peterson, Richard E., and Nabil El-Ramly, "The Worldwide Electric and Nonelectric Geothermal Industry." Paper from Hawaii Natural Energy Institute, 1975.

Phelps, Paul L., and Lynn R. Anspaugh, *Imperial Valley Environmental Project: Progress Report,* UCRL-50044-76-1, Lawrence Livermore Laboratory, California, 1976.

Pick, James B., and Edgar W. Butler, "The Sociological Impact of Geothermal Development," in Edmunds, Stahrl, and Adam Rose (eds.), *Geothermal Energy and Resource Development,* New York: Praeger, 1979.

Pick, James B., Tae Hwan Jung, and Edgar W. Butler, "Regional Employment Implications For Geothermal Energy Development, Imperial County, California." *Proceedings of the 3rd Los Angeles Area Energy Symposium.* Los Angeles, Western Periodicals, pp. 159–167, 1977.

Pick, James B., Tae Hwan Jung, and Edgar W. Butler, *Population Analysis Relative to Geothermal Energy Development, Imperial County, California,* Dry Lands Research Institute, Riverside, California (DLRI Report No. 16), 1977.

Pick, James B., Charles Starnes, Tae Hwan Jung, and Edgar W. Butler, "Population Economic Data Analysis Relative to Geothermal Fields, Imperial County, California." *Proceedings of the American Statistical Association, Social Statistics Section,* **Vol. II**: 666–672, 1976.

Pick, James B., Charles Starnes, Tae Hwan Jung, and Edgar W. Butler, "Correlates of Fertility and Morality in Low Migration Standard Metropolitan Statistical Areas," *Social Biology* **24**(1): 666–672, 1977.

Pick, James B., "Computer Display of Population Age Structure," *Demography* **11**(4): 673–682, 1974.

References

Pierson, D., in *Proceedings of the 11th Intersociety Energy Conversion Engineering Conference*, American Institute of Chemical Engineers, New York, pp. 850–852, 1976.

Pimental, David, William Dritschelo, John Krummel, and John Kutzman, "Energy and Land Constraints in Food Protein Production," *Science* **190**: 754–761, 1975.

Pivirotto, D. S., in *Proceedings of the 11th Intersociety Energy Conversion Engineering Conference*, American Institute of Chemical Engineers, New York, pp. 843–849, 1976.

Price, John A. *Tijuana: Urbanization in a Border Culture*. Notre Dame, Indianapolis: University of Notre Dame Press, 1973.

Raleigh, C. B., J. H. Healy, J. D. Bredehoeft, "An Experiment in Earthquake Control at Rangely, Colorado," *Science* **191**: 1230–1237, March, 1976.

Reed, M. J., and G. E. Campbell, "Environmental Impact of Development in the Geysers Geothermal Field, USA," in *Proceedings of the 2nd United Nations Symposium on the Development and Use of Geothermal Resources*, U.S. Government Printing Office, Washington, D.C., pp. 1399–1410, 1976.

Renner, J. L., D. E. White, and D. L. Williams, "Hydrothermal Convection Systems," in *Assessment of Geothermal Resources of the U.S.—1975*, USGS Circular 726, 1975.

Rex, Robert W., "The U.S. Geothermal Industry in 1978," *Geothermal Energy* Vol. 6, No. 7, pp. 28–37, 1978.

Rex, Robert W., "Geothermal Resources in the Imperial Valley," in Seckler, David (ed.), *California Water*, Berkeley, University of California Press, 1971.

Richards, Richard G., "A Utility's Evaluation of Potential Broad-based Geothermal Development." *Proceedings of a Conference on the Commercialization of Geothermal Resources*. Davis, California: Geothermal Resources Council, 1978.

Robson, Geoffrey R., "Geothermal Electricity Production," *Science* **84**: 371–375, 1974.

Rodzianko, Paul, "Overview of and Introduction to Geothermal Power Plant Financing Options," *Proceedings of a Conference on the Commercialization of Geothermal Resources*. Davis, California: Geothermal Resources Council, pp. 35–38, 1978.

Rogers, H. I., "Financing and Construction of a Commercial Geothermal Power Plant in Utah." *Proceedings of a Conference on the Commercialization of Geothermal Resources*. Davis, California: Geothermal Resources Council, pp. 39–42, 1978.

Rose, Adam, *The Economic Impact of Geothermal Energy Development*. Riverside: University of California, Dry Lands Research Institute, 1977.

Ross, S. H., "Geothermal Potential of Idaho," *Geothermics* Special Issue 2, 1970.

Saben, Samuel, "Geographic Mobility and Employment Status, March 1962–March 1963," *Monthly Labor Review*, 1964.

Sacarto, Douglas M., *State Policies for Geothermal Development*. Denver, Colorado. National Conference of State Legislatures, 1976.

Samora, Julian, *Los Mojados: The Wetback Story*. Notre Dame, Indianapolis: University of Notre Dame Press, 1971.

Science Applications, Inc., *Regional Operations Research for Geothermal Energy Resources in California and Hawaii*. Technical Progress Report (3.B.b.). La Jolla, California: Science Applications, Inc., 1978.

Schultz, Robert J., Joseph A. Hanny, and William H. Knuth, "Geothermal Energy Market Potential in Industrial Processing." *Proceedings of a Conference on the Commercialization of Geothermal Resources*. Davis, California: Geothermal Resources Council, pp. 27–30, 1978.

Semrau, Konrad T., "Control of Hydrogen Sulfide From Geothermal Power Production," in *Geothermal Environmental Seminar—1976*, pp. 185–189, Shearer-Graphic Arts, Lakeport, California, 1976.

Sharp, R. P. *Geology Field Guide to Southern California*. Dubuque, Iowa: William C. Brown Company, 1972.

Sheehan, Mike et al. Report of the Estimated Impact of Locating a One Hundred Acre Geothermal Facility on Various Land. Preliminary report, NSF-Imperial County Geothermal Project, August, 1976.

Sherwood, Peter B., "The Economic Feasibility of Utilizing Geothermal Heat for an Agricultural Chemical Plant," in Direct Utilization of Geothermal Energy: A Symposium. Davis, California: Geothermal Resources Council, pp. 91–95, 1978.

Shinn, J. H., B. R. Clegg, M. L. Stuart, and S. E. Thompson, "Exposures of Field-Grown Lettuce to Geothermal Air Pollution—Photosynthetic and Stomatal Responses." Preprint UCRL-78238, Lawrence Livermore Laboratory, 1976.

Shively, R. W., "Decision-Making for Locating Industry," in Whiting, Larry R. (ed.), *Rural Industrialization*. Ames: Iowa University Press, 1974.

Shryock, Henry S., Jacob S. Siegel and Associates, Condensed edition by Edward G. Stockwell. *The Methods and Materials of Demography*. New York: Academic Press, 1976.

Southern California Edison Company, 1977 Annual Report. Rosemead: Southern California Edison Company, 1977.

State of California, *California Statistical Abstract, 1975*. Sacramento: State Printing Office, 1976.

State of California, Department of Public Health. *Vital Statistics of California*. Various volumes. Sacramento, California, 1930–1976.

State of California, Department of Public Health. Personal communication, December, 1967a.

Steinhart, John S., and Carol E. Steinhart, "Energy Use in the U.S. Food System," *Science* **184**: 307–316, 1974.

Stilwell, W. B., W. K. Hall, and J. Tawhai, "Ground Movement in New Zealand Geothermal Fields," in *Proceedings of the 2nd United Nations Symposium of the Development and Use of Geothermal Resources*, U.S. Government Printing Office, Washington, D.C., pp. 1427–1434, 1975.

Stoddard, Ellwyn R., "Illegal Mexican Labor in the Borderlands," *Pacific Sociological Review* **19**(12): 175–210, 1976.

Sweet, James A., "The Employment of Rural Farm Wives," *Rural Sociology*, **37**(4): 553–577, 1972.

Ternes, Mathais Nicholas, *The Public Interest and Geothermal Energy in Heber, California*. Master's thesis, University of Oklahoma, Norman, Oklahoma, University of Oklahoma, 1978.

Thompson, Ray, "Behavior of H_2S in the Atmosphere and Its Effect on Vegetation," in Geothermal Environmental Seminar—1976, Fayne L. Tucker and Mary Dean Anderson (eds.), pp. 193–198, Shearer-Graphic Arts, Lakeport, California.

Thompson, Warren S., *Growth and Changes in California's Population*. Los Angeles: Haynes Foundation, 1955.

Towse, D., "An Estimate of the Geothermal Energy Resource in the Salton Trough, California." Report No. UCRL-51851, Lawrence Livermore Laboratory, University of California, 1976.

Transmission Route Selection Committee, Imperial County, Committee Proceedings. El Centro, California: Transmission Route Selection Committee, 1977.

Trehan, R., A. Cohen, J. Gupta, W. Jacobsen, J. Leigh, and S. True, *Site-Specific Analysis of Geothermal Development*. Vol. II, Scenarios and Requirements, MITRE Corporation (MTR-7586), Washington, D.C., U.S. Department of Energy, August, 1978.

TRW Inc., *Energy Transport Costs*. Unpublished report. Redondo Beach, California, TRW Inc., 1977a.

References

TRW Inc., *Use of Geothermal Heat for Sugar Refining*. Final Report, ERDA Contract No. E(04-3)-1317. Redondo Beach, California, TRW Inc., 1977b.

Turk, Herman, "Interorganizational Networks in Urban Society: Initial Perspectives and Comparative Research," *American Sociological Review* 35(February, 1970): 1–19.

Uhlenberg, Peter, "Fertility Patterns Within the Mexican-American Population," *Social Biology* 20(1): 30–39, 1973.

U.S. Bureau of the Census. Census of Population. Various volumes. Washington, D.C., Government Printing Office, 1910–1970.

U.S. Bureau of the Census, *U.S. Census of Population, 1970*. Fifth Count for ZIP Codes, Counties, and Smaller Areas. Washington, D.C.: Government Printing Office, 1974.

U.S. Bureau of the Census, *U.S. Census of Population, 1970*. General Population Characteristics. Final Report PC(1)-B. Various volumes. Washington, D.C. Government Printing Office, 1973.

U.S. Bureau of the Census, Census of Housing: 1970. Vol. 1, Housing Characteristics for States, Cities and Counties. Washington, D.C.: Government Printing Office, 1971.

U.S. Bureau of the Census, *Census of Population: 1970*. General Social and Economic Characteristics. Final Report PC(1)-C. Washington, D.C.: Government Printing Office. 1971.

U.S. Bureau of the Census, *Census of Population: 1960*. General Social and Economic Characteristics. Final Report PC(1)-C. Washington, D.C.: Government Printing Office, 1961.

U.S. Bureau of the Census, *Census of Population: 1950*. Vol. 2 Characteristics of the Population. Washington, D.C.: Government Printing Office, 1951.

U.S. Department of the Interior, Geological Survey, *Assessment of Geothermal Resources of the United States—1978*, USGS Circular 790, L. J. P. Muffler, Editor, 1979.

U.S. Department of the Interior, Geological Survey, *Assessment of Geothermal Resources of the United States—1975*, USGS Circular 726, D. E. White and D. L, Williams, Editors, 1975.

U.S. Department of the Interior and the Resources Agency of California, Salton Sea Project, California, April, 1974.

U.S. Nuclear Regulatory Commission and U.S. Department of the Interior, Site Environmental Statement Relative to Determination of the Suitability of the Proposed Site for Eventual Construction of Sundesert Nuclear Plant Units 1 and 2. Docket Nos. 50-582 and 50-583. Washington, D.C.: U.S. Nuclear Regulatory Commission, 1978.

U.S. Senate, 91st Congress, Migrant and Seasonal Farmworker Powerlessness. Hearings on Border Commuter Problem. Washington, D.C.: Government Printing Office, 1971.

Van Arsdol, Maurice D., and Leo A. Schuerman, "Redistribution and Assimilation of Ethnic Populations: The Los Angeles Case," *Demography* 8(4): 459–480, 1971.

Verdugo, Sergio N., "El Nivel de Vida Regional en Mexico," *Boletin Economico* 1(1): 26–48, 1974.

Vollintine, Larry, and Oleh Weres, *Public Opinion Concerning Geothermal Development in Lake County, California*. Berkeley, California: Lawrence Berkeley Laboratory, March, 1976.

Vollintine, Larry, and Oleh Weres, *Public Opinion in Cobb Valley Concerning Geothermal Development in Lake County, California*. Berkeley, California: Lawrence Berkeley Laboratory, June, 1976.

VTN Inc., Untitled water report. 113 pages. Irvine, California, VTN Inc., 1978.

Wagner, Sharon, "A Review of Western State Tax Law Applied to Geothermal Development." *Proceedings of a Conference on the Commercialization of Geothermal Resources,* pp. 65–68, 1978.

Walton, John, "The Vertical Axis of Community Organization and the Structure of Power," *Social Science Quarterly* **48**(December, 1967): 353–368.

Warren, Roland L., "Toward a Typology of Extra-Community Controls Limiting Local Community Autonomy," *Social Forces* **34**(May, 1956): 338–341.

Weber, Max. *The Theory of Social and Economic Organization,* translated by A. M. Henderson and Talcott Parsons, Glencoe, Illinois: The Free Press, 1957.

Weres, Oleh, "Environmental Implications of Exploitation of Geothermal Brines," in *Geothermal Environmental Seminar—1976,* Fayne L. Tucker and Mary Dean Anderson (eds.), pp. 115–123, Shearer-Graphic Arts, Lakeport, California.

Wehlage, Edward F. "Geothermal Energy's Potential for Heating and Cooling in Food Processing," in Lienau, Paul J., and John Lund (eds.), *Multipurpose Use of Geothermal Energy.* Klamath Falls: Oregon Institute of Technology, 1974.

Wehlage, Edward F., *The Basics of Applied Geothermal Engineering.* West Covina, California: Geothermal Information Services, 1976.

Werner, H. H., "Contribution to the Mineral Extraction from Supersaturated Geothermal Brines, Salton Sea Area, California," *Geothermics* Special Issue 2, 1970.

Whetten, Nathan L. *Population Growth in Mexico.* Paper prepared for the Select Commission on Western Hemisphere Immigration, 1971.

White, D. E., and D. L. Williams (eds.), *Assessment of Geothermal Resources of the United States—1975.* Menlo Park, California: U.S. Geological Survey, 1975.

Whiting, Larry R. (ed.), *Rural Industrialization* Ames: Iowa State University Press, 1974.

Williams, F., A. Cohen, R. Pfundstein, and S. Pond, *Site Specific Analysis of Geothermal Development.* Vol. III, Data Files of Prospective Sites, MITRE Corporation, Washington, D.C., 1978.

Williams, F., A. Cohen, R. Pfundstein, and S. Pond, *Site Specific Analysis of Geothermal Development—Data Files of Prospective Sites, Vol. III.* HEP/T4014-01/3, UC-66. Washington, D.C., 1978.

Wilson, R. D., "Use of Geothermal Energy at Tasman Pulp and Paper Co. Ltd., New Zealand," in Lienau, Paul J., and John Lund (eds.), *Multipurpose Use of Geothermal Energy.* Klamath Falls: Oregon Institute of Technology, 1974.

Wong, C. M., "Geothermal Energy and Desalinization: Partners in Progress," *Geothermics* Special Issue 2, 1970.

Woods, John H., "Structuring of Geothermal Development Loans." *Proceedings of a Conference on the Commercialization of Geothermal Resources.* Davis, California: Geothermal Resources Council, pp. 51–54, 1978.

Wright, Philip M., "Nature, Occurrence and Utilization of Geothermal Energy." *Proceedings of a Conference on the Commercialization of Geothermal Resources.* Davis, California: Geothermal Resources Council, pp. 1–10, 1978.

Wyllie, Peter J., *The Way The Earth Works.* New York: John Wiley & Sons, 1976.

Subject Index

Age, eligibility, 145
Age structure, 91-98
Agriculture, 5, 313
　revenue, 6
　workers, 13
Air pollution, 307-308
Akureyri, Iceland, 278
Alamo River, 7, 67, 72, 296-297
All American Canal, 6, 8, 67, 262, 296
American Association of University Women, 259
Anderson, J. Hilbert, Inc., 264
Arizona, 253, 291
Asthenosphere, 36
Atlantic Ocean, 35, 36, 37, 62

Baja, California, 8-9
Binary power plant, 50-52, 264-266, 297
Binary test plant—Magma Power, 264-266
Blowouts, 312
Boom towns, 323
Border, commuters, 140, 166
　industrialization program, 88
　regulations, 315
Brady Hot Springs, 309
Brawley, California, 5, 10, 85, 107, 115, 123, 250-252, 257, 275
　KGRA, 268, 271, 281
　sewage plant dispute, 251-252
Brine waste disposal, 303
Bureau of Land Management, 259-262, 288

Calexico, California, 5, 10, 79, 87-88, 89, 107, 111, 112, 115, 123, 257, 262
California State Department of Parks and Recreation, 260

California State Department of Wildlife, 262
California State Division of Oil and Gas, 287, 288, 289
California State Energy Commission, 255, 288
California State Public Utilities Commission, 257, 288, 291
Calipatria, California, 257
Cerro Prieto, Mexico, *vi*, 47, 55, 303, 307
CETA, 252
Chevron, 251, 267, 268, 276
Chicanos, *see* Spanish Americans
Citizen involvement, 323-324
Coachella Canal, 296
Coal, 55, 255, 290
Coco Hot Springs, 309
Colorado River water, 5-8, 67, 71, 296, 309
Comptroller General's Office (U.S.), 199-200
Computer mapping, 107
　dependency ratio, 107, 109
　female labor force, 115, 117
　income, 118
　male labor force, 112-116
　mobility, 116, 118-121
Conflict, agribusiness and geothermal development, 324
Consolidation, 17
Continental drift theory, 36
Cooling towers, 70, 298, 300
Cooling water, 50, 70-71, 266, 296-301
County of Imperial, *see* Imperial County

Data analytic techniques, 26-27
Deep Sea Drilling Project (NSF), 36

Department of Energy (DOE), *xi,* 43, 252, 267, 271, 275, 278, 292
Developer
  IRAC contract, 293
  utility contract, 293
Diesel power, 89
Direct use, 53, 74, 79, 88, 175-176, 217-274, 317
  industrial processes, 272-273
  retrofit problem, 274
  temperatures, 272
  transmission pipes, 295-296
  (*See also* Nonelectrical use)
Discriminant analysis, 27, 122-128
  explanations of, 122, 123
  implications of, 128, 129
Drilling towers, 43, 44, 45
  development, 21, 22
  exploration, 21-22

Earthquakes, 36, 37, 68
East Mesa, *vi, ix,* 107, 259-261, 293, 297, 306
El Centro, California, 5, 107, 115, 116, 223, 257, 271, 274-279, 282
El Centro Community Center, 271, 274-279
El Centro Naval Air Station, 262
Electrical exploratory techniques, 61
Electrical use developments, 264-271
Electric Power Research Institute, 267
Elsinore fault, 40
Employers, 14
Enhancement, 18
Enumeration district (ED), 78, 107
Evaporation pond, 47, 72
Environmental impact statement, 288, 306-313, 327-328

Family composition, 99
Farm labor, 161-162, 165-172
  reduction, 172-175, 315
Federal Loan Guarantee Program, 262
Field crops, 6
Field development costs, 291
Financing, 290-294
Fiscal impact of geothermal development, 187-193
Flash power plant, 47-48
Flash test plant, Union Oil, 268
Fuel mix, Imperial County, 89

Gas emissions, 49, 50
General plan, Imperial County, 30-31, 252, 296, 305 ff., 317-319
Geochemical exploration techniques, 45
Geological exploration techniques, 45
Geopressure, 43
Geothermal Element, Imperial County General Plan, 252, 296
Geothermal Loan Guarantee Program, 292
Geothermal power plants, 46-53 (*see* Power plant)
Geothermal resources
  conduction dominated, 43
  costs of, 56, 291
  definitions, 35
  energy, 35
  exploration and drilling, 43-46, 60-61
  geological theory, 35-40
  hot igneous rock, 42-43
  hydrothermal, 40-41
  impact on leadership, 251-263
  impact on population, 250-263
  Imperial County, 60
  life expectancy, 68-69
  origin, 62-64
  ownership of, 69
  power plants, 46-53
  radiogenic, 43
  types, 40-43
  use of, 58
  worldwide generating capacity, 59
Geothermal Steam Act of 1970, 288
Geysers, (*see* The Geysers)
Gravimeter, 43
Gravity explorations techniques, 43-44, 60
Ground and aeromagnetic exploratory techniques, 61
Groundwater, 66-67, 71, 296-297

Heat exchanger, 266, 276
Heber, *vi, ix,* 84, 107, 282
  demonstration plant proposal, 267-268
  KGRA, 170, 267, 276, 281, 282, 328
  public opinion, 230-232
Himalayas, 36
Holly Sugar plant, 271, 279-282
Holtville, 257
Hot dry rock, 42
Hot igneous rock, 42
HUD, 275, 279
Hydroelectric power, 89

Subject Index

Iceland, 37
Identification, 18
IID (*see* Imperial Irrigation District)
Immigrants, illegal, 9-10, 85, 129, 130, 133-136
  legal, 9-10, 85, 129
Imperial (city), 257
Imperial County, *v*, 79-80, 216, 261
  age structure, 91-100
  energy capacity and consumption, 88, 139
  energy consumption, 89
  energy planning, 31-34
  Farm Bureau, 257, 261, 262
  general plan (*see* General Plan)
  geothermal areas, 66-67
  history, 79-80
  industrialization, 177
  orientation, 3-14
  planning, 31-34
  Planning Commission, 257
  Planning Department, 262
  population, 5, 8-14, 89-101
  research perspective, 30
  water allocation, 297
Imperial Irrigation District (IID), 6, 253, 259, 260-261, 262, 290
Imperial Valley Action Plan, 253, 257, 291
Imperial Valley College, 252
Industrial categories, 13
Interim Risk Assuming Company, 293
International Brotherhood of Electrical Workers, 252
Interstitial land reduction (ILR), 168, 170, 172
Irrigation canals, 50, 53

Jet Propulsion Laboratory, 255

KGRA (Known Geothermal Resource Area), 79
Klamath Falls, Oregon, 287

Labor force participation, 146
Labor interaction, Calexico and Mexicali, 87-88
Lake County, California, 77, 79
Land acquisition, 21
Land consumption from geothermal development, 73-74
Land subsidence (*see* Subsidence)

Lardarello, Italy, 77
Lawrence Livermore Laboratory, 305
Lead agency, 288
Leadership, 235 ff.
  disputes, 251-263, 294
  effect on geothermal, 244-246
  influential people, 237-240
  opinion, 241-244
  opinion about geothermal, 240-241
  opinion on public, 241-244
  power structure methodology, 236-237
Leveraged lease, 293
Limiting factors, geothermal development, 289-303
Lithosphere, 36
Los Alamos, New Mexico, 42
Los Angeles, California, *vi*, 253

Magma, 42
Magma Power Company, 251, 264, 270, 271, 292
Magnetic anomaly, 45
Make-up water, 50, 297
Methodology, 19-20
  data collection, 20
Mexicali, Mexico, 8, 79, 84, 87, 166, 296
Mexicali Valley, 67
Mexican-American (*see* Spanish Americans)
Mexico, 8-9, 83
Migration, 80-81, 86-87, 98-99, 116, 118, 162, 178-179, 196, 313-314
Mitigation, 18-19
Mono Lake–Long Valley, 309
Mono Power Subsidiary, SCE, 270
Morrison Knudsen Company, 271
Multipurpose use, 283

National Parachute Test Range—U.S. Navy, 259, 260, 262
National Science Foundation, *xi*
New River, 7, 67, 72, 251, 268, 296-297
New Zealand, *vi*
Niland, California, *vi*, 116, 264, 271
Noise levels, 49
Nonelectrical use (*see* Direct use)
North Brawley, *vi, ix*
Northern Baja California, Mexico, 83
Nuclear energy, 255, 291

Ocotillo, California, 257

Subject Index

Pacific Ocean, 62-63
Pangaea, 36
Permeability, 41, 45
Phoenix, Arizona, 257
Pilot test facility, SDG & E, 270
Planning, research oriented, 330-331
Plate tectonics (see Tectonic plates)
Policy
 alternatives, 26-29
 federal, 21-22
 outcomes, 26-29
 population analysis, 318-319
 priorities, 26-29
 public opinion, 199-200, 323, 324
 state, 21-22
Political limiting factors, 294
Political processes, 25-26
Population
 change, 9-10
 economics of Imperial County, 100-103
 Imperial County, 80-81
 income, 100-101
 industrial distributions, 101-102
 labor force, 103
 projections, 165, 187, 316, 381
  assumptions of, 179-183
  farm labor, 166-172
  methods for, 176, 196
  results, 183-187
Porosity, 41
Power plant, 46-53, 72, 73, 291, 297-301
 capacity additions, 181, 183, 255
 costs, 291
 siting, 72-73
Power structures (see Leadership)
Price, geothermal energy, 54
Public opinion survey, 199 ff., 322
 about geothermal development, 205-210
 adequacy of information, 216-218
 attitudes, 211-216
 cautions, 205
 conclusions, and recommendations, 232, 233
 future research, 233-234
 Heber KGRA, 230-232
 landowners, 226-228
 leadership, 241-244
 mail-back procedures, 204-205
 methodology, 201-204

Public opinion survey (continued)
 problems, 221-225
 questionnaire design and review, 204
 regulation, 219-221
 response rates, 205
 rural landowners, 228-230
 transferability, 233
 understanding of geothermal development, 210-211

Regional Water Quality Board, 252, 288
Regression analysis, 141-161
 explanations of, 161-164
 implications of, 161-164
 results of, 152-161
 sources of error, 196
Regulations, 56-58, 72-73, 287-289
Regulatory permitting process, 287-289
Reinjection, 67, 265
Republic Geothermal Corporation, 107, 251, 271, 293
Research results
 assumptions, 324-327
 conclusions, 306 ff.
 model, 16, 17-20
 population and economics, 313-318
 public opinion, 319-324
 transferability, 2
Reservoir
 capacity estimates, 64-65, 69
 engineering, 45-46, 67
 insurance, 294
 recharge, 67
Reykjavik, Iceland, 296
Risks, geothermal financing, 291
Riverside County, California, 255
Rogers Engineering, 268
Roosevelt Hot Springs, 309
Rural industrialization, 86-87

Salinity, 66, 67, 71, 264, 267, 268, 270, 297
Salton City, California, 257
Salton Sea, ix, 271, 296, 306, 310-312
 KGRA, 60, 65-66, 67, 270, 297
 recreation, 311
Salton Trough, 3
San Andreas Fault, 36, 40, 62
San Diego, California, vi, 257, 291

Subject Index

San Diego Gas & Electric Company (SDG & E), 252, 255, 262, 264, 270, 271, 292
San Jacinto Fault, 40
Sea floor spreading, 36
Seasonality, 195
Seismic exploratory techniques, 44, 61
Seismometer, 44
Sex ratio, 80
Shared projects, 294
Slant drilling, 169-170, 265
Socioeconomic impact, 328-329
Sonoma County, California, 19, 77
South America, 36
South Brawley, *ix*
Southern California Edison Company (SCE), 252, 268, 292
Southern Pacific Land Company, 270
Spanish Americans, 7, 10, 12-14, 81-86, 313-315
  fertility, 131
  geographic distribution, 124
  population, 129-133
Standard Metropolitan Statistical Area, 195
Subsidence, 72, 170, 297, 306-307
Sun Desert nuclear plant, 255, 257
Surface manifestation of geothermal resources, 43, 66
Systematization, 18

Taxation, 249, 292
Taxation laws, 249
Tectonic plates, 35-40, 62-63
Temperature
  estimation, 64
  gradient, 44, 61
  reservoirs, 66, 264, 267, 268, 270
Test wells, 45, 60, 61

The Geysers, *vi*, 2, 19, 50, 56, 68, 165, 176, 181, 307-309
Tile drains, 169
Transferability of methods and results, 194-197
Transform faults, 37-38
Transmission
  corridor routes, 256, 258, 262, 263
  Corridor Selection Committee, 256, 257
  lines, 194, 253-263, 290
  pipes, 295-296
  pipes cost, 295-296
  towers, 253
Transportation, 106
Transport of hot water, 175, 276-278, 295-296
TRW, Inc., 279

Unemployment, 145, 314, 318
Union Oil Company, 250, 268, 270
Union workers, 252
University of California, Riverside, *ix*
U.S. Geological Survey (USGS), 23, 288
U.S. Weather Service, 22-23

Valle Calderas, New Mexico, geothermal field, 268
Valley Nitrogen fertilizer plant, 271, 282-287
Volcanoes, 36
VTN, Inc., 297

Wairakei, New Zealand, 47, 303, 307
Water supply, 50, 70-72
Well pump, 266
Westec Services, 250, 275
West Mesa, 107, 259-261
Westmoreland, California, *ix*, 259

Yuma, Arizona, 296

Author Index

Allen, W., 307, 341
Alvirez, D., 345
Anderson, David N., 2, 341
Anderson, M., 342
Anspaugh, Lynn R., 305, 348
Armstead, C., 341
Armstrong, DeVonne W., 233, 341
Avery, R., 122
Axtmann, R. C., 307, 341

Baer, R., 145, 147, 148, 153, 341
Baker, J., 341
Barnea, Joseph, 88, 341
Beale, C., 186, 341
Bean, F., 342
Biehler, Shawn, *xii*, 60, 63, 64, 66-69, 75, 341, 343
Biggar, J. C., 145, 147, 341
Birsic, R., 341
Bogue, D., 179, 181, 341
Boughey, A., 341
Boyton, J., 148, 345
Bradshaw, B., 82, 90, 131, 342
Bredhoeft, J. D., 349
Brugram, J., 345
Buck, J. Vincent, 7, 342
Buddy, D., 147, 345
Bufe, C. G., 6, 68, 342
Butler, Edgar W., *vi, vii, xii*, 118, 137, 147, 227, 233, 335, 341, 342, 348

Campbell, G. E., 307, 349
Castrantas, Harry M., 307, 342
Cataldi, R., 342
Chamberlain, L. L., 346
Chapin, F. Stuart, 342
Chasteen, A. J., 312, 342
City of El Centro, 277, 279, 280, 281
Clegg, B. R., 350
Cohen, A., 346, 350, 352
Collver, O. A., 145, 146, 342
Comptroller General, 342
Cortese, Charles, 6, 346

Dagodog, W. Tim, 84, 130, 342
Dambly, Benjamin W., 265, 342
Davis, Cal, 168, 172, 180, 183, 315-316, 343
DeSandre, P., 147, 343
Dibblee, T. W., Jr., 3, 343
Defferding, Leo J., 303, 343
Dillman, Don A., 234, 343
DiMario, P., 342
Dixon, W. J., 151, 343
Draper, N. R., 147, 343
Dritschelo, William, 348
Dunlap, Riley E., 234, 343

Easterlin, R. A., 196, 343
Edmunds, Stahrl, 343, 347
Eisenbeis, Robert A., 122, 343

Eisenstat, Samuel M., 343
Elders, Wilfred A., *xii*, 3, 4, 37, 39, 40, 63, 343
Electric Power Research Institute, 267
El-Ramly, Nabil, 348
Environmental Systems Research, 107, 343
Ermack, Donald L., 69-71, 73, 74, 166, 185, 305, 343, 346
Estrada, Leo, 345
Evans, D. M., 343

Facca, Giancarlo, 59, 347
Falk, Harry W., Jr., 345
Farnan, Maureen, 226, 227, 239-240, 344
Finnemore, E. John, 306, 343
Flieger, Wilhelm, 179, 346
Force, Peter R., 88, 343
Fowler, John M., 79, 88, 89, 192, 327, 343
Freudenberg, William R., 218, 343
Fulton, Maurice, 87, 141, 343

Galle, Omer, R., 125, 145, 147, 343, 344
Gelbner, Christopher G., 344
Geothermal Hot Line, 344
Geothermal Resources Council, 344
Geyser, The, 344

Author Index

Ginther, Jerry D., 86, 344
Goldberg, Kalman, 235, 344
Goldman, G., 187, 190, 191, 193, 244, 344
Goldsmith, Martin, 295, 296, 301, 302, 310, 344
Gormely, Patrick J., 344
Gove, Walter R., 343, 344
Gray, Irwin, 149, 150, 344
Grebler, Leo, 82, 127, 130, 131, 344
Green, Phyllis, 226, 227, 239-240, 344
Greenwood, Michael J., 344
Griscom, T., 61, 344
Grupta, J., 350
Gusman, Ralph G., 344

Hall, Charles, *xii*, 248
Hall, W. K., 350
Handler, Phillip, 196, 345
Hanny, Joseph A., 349
Harbon, Don C., 345
Haren, Claud, 345
Hauser, Phillip, 87, 345
Hawley, Amos H., 235, 345
Healy, J. H., 349
Heer, D. M., 145, 147, 148, 345
Hemmens, George C., 342
Hernandez, Jose, 82, 90, 345
Hernandez, Juan P., 345
Hill, D. P., 307, 345
Hinrichs, Thomas C., 345
Holt, Ben, 345
Hornberger, G. A., 344
Hunter, L. C., 147, 345

Icerman, L., 89, 348
Imperial County, 345
Imperial Valley Action Plan, Utility Technical Committee, 259, 345

Jacobsen, W., 346, 350
James, R., 88, 345

Jessen, Raymond J., 203, 345
Jet Propulsion Laboratory, 345
Johnson, Claude, 161, 165, 168, 228, 316, 345-346
Johnston, David, 346
Jones, Bernie, 346
Jung Tae Hwan, *xii*, 137, 348

Kaiser, Edward, 342
Kaplan, Abraham, 235, 346
Kelly, R. E., 308, 346
Keyfitz, Nathan, 179, 196, 346
Kjos, Kaare Saxe, 79, 83, 84, 87, 140, 346
Knuth, William H., 349
Krummel, John, 348
Kuhn, Philip H., 177, 346
Kutzman, John, 348

Lachenbruch, Peter, 122, 127, 196, 346
Langlois, E., 147, 148, 342
Lansing, John B., 81, 346
Lapp, Ralph, 346
Larson, Tod, 346
Lasker, R., 301, 346
Lasswell, Harold D., 235, 346
Layton, D. W., 69-72, 166, 244-245, 246, 305, 346
Leardini, T., 342
Lee, Tien-Chang, 60, 63, 64, 66-69, 75, 307, 341, 346
Leigh, J. A., 17, 346, 350
Leitner, P., 49, 346
Lessing, Lawrence, 346
Lester, F., 68, 342
Levine, Louis, 145, 346
Lienau, Paul J., 272, 274, 347, 352
Lindal, B., 88, 347
Lindow, Howard E., 344
Lindsay, Donald R., 287, 347

Lindstromberg, Robin, 235, 344
Linton, A. M., 347
Lofgren, B. E., 306, 347
Lofting, Everard, 140, 165, 190, 235, 236, 244
Lund, John W., 347, 352
Lundy, Robert, *xii*

Magma Power Company, 347
May, R. C., 301, 347
McCarthy, John D., 343, 344
McCluer, H. K., 307, 341
Meidav, H. Tsvi, 38, 59, 343, 347
Meidav, Mae Z., 347
Melvin-Howe, G., 147, 347
Moore, Joan W., 344
Morrison, Peter A., 81, 347
Mowinchel, P., 345
Muffler, L. J. P., 2, 22, 61, 344, 347
Mueller, Eva, 81, 346
Myers, George C., 177, 347

Nakamura, S., 347
Nathan Associates, 83, 347
National Petroleum Council, 347

Otte, Carl, 291, 312, 348

Palmer, T. D., 343
Panawek, Gregory, 293, 348
Pasqualetti, Martin, *xii*, 34, 181, 331, 348
Paulson, Wayne, 348
Peake, L. G., 345
Pearson, Russ, 348
Penner, S. S., 348
Perry, G., 348
Peterson, Richard E., 348
Pfundstein, R., 352
Phelps, Paul L., 305, 348

Author Index

Pick, James, *vi, vii, xii*, 118, 124, 137, 145, 146, 147, 165, 175, 196, 244, 341, 342, 348
Pierson, D., 324, 348
Pimental, David, 75, 348
Pivirotto, D. S., 348
Pond, S., 352
Pope, Hallowell, 348
Price, John A., 349

Raleigh, C. B., 307, 349
Reed, M. J., 307, 349
Reid, G. L., 147, 345
Renner, J. L., 65, 349
Rex, Robert W., V, *ix, x, xii*, 2, 168, 342, 343, 349
Richards, Richard G., 291, 294, 349
Robinson, P. T., 343
Robson, Geoffrey, R., 349
Rodzianko, Paul, 293, 349
Rogers, H. I., 293, 349
Rose, Adam, 140, 150, 165, 168, 172, 177, 343, 347, 349
Ross, S. H., 349

Saben, Samuel, 81, 349
Sacarto, Douglas M., 54, 55, 57, 58, 59, 349
Samora, Julian, 84, 130, 135, 136, 166, 315, 349
Sanyal, S., 59, 347
Schuerman, Leo A., 351
Schultz, Robert J., 272, 273, 349
Science Applications, Inc., 349
Semrau, Konrad T., 307, 349
Sharp, R. P., 3, 349
Sheehan, Mike, 165, 168, 172, 349

Sherwood, Judith, 345
Sherwood, Peter B., 283, 284, 286, 349
Shinn, J. H., 307, 350
Shively, R. W., 315, 344, 350
Shyrock, Henry S., 176, 181, 350
Siegel, Jacob S., 176, 181, 350
Smith, H., 147, 343
Southern California Edison Co., 350
Starnes, Charles, 137, 348
State of California, 180, 350
Stegman, Michael A., 342
Steinhart, John S., 88, 350
Stilwell, W. B., 307, 350
Stoddard, Ellwyn R., 85, 135, 350
Strong, D., 187, 190, 191, 193, 244, 344
Stuart, M. L., 350
Sweet, James A., 145, 147, 148, 350
Sweetwood, Douglas, 344

Tawhai, J., 350
Tenaza, R. H., 346
Ternes, Mathais Nicholas, 56, 166, 193, 231-232, 350
Thompson, L. R., 350
Thompson, Ray, 307, 350
Thompson, Warren S., 80, 350
Towse, D., 65, 350
Transmission Rate Selection Committee, Imperial County, 256, 257, 350
Trehan, R., 17, 346, 350
True, S., 350
TRW, 295, 350

Tucker, Fayne L., 342
Turk, Herman, 351
Turner, Thomas, A., 342, 345

Uhlenberg, Peter, 82, 131, 351
U.S. Senate, 83, 85, 132, 166, 351

Van Arsdol, Maurice D., 351
Verdugo, Sergio N., 351
Veysey, Victor V., *v-vii, xii*
Vollintine, Larry, 77, 78, 79, 204, 205, 219, 351
VTN, 48, 51, 52, 297, 299, 300, 351

Wagner, Sharon, 292, 351-352
Walton, John, 235, 352
Warren, Roland L., 352
Weber, Max, 237, 352
Wehlage, Edward F., 77, 88, 352
Weiss, Shirley F., 142
Weres, Oleh, 77, 78, 79, 204, 205, 219, 351, 352
Werner, H. H., 88, 352
Whetten, Nathan L., 8, 83, 352
White, D. E., 2, 22, 352
Wilkin, D., 341
Williams, D. L., 2, 17, 22, 352
Williams, F., 352
Wilson, R. D., 352
Wong, C. M., 352
Woods, John H., 352
Wright, Philip M., 40, 352
Wyllie, Peter J., 36, 37, 38, 352